GEOGRAPHY OF A HIMALAYAN KINGDOM : BHUTAN

About the Authors

Neil Fraser studied at Oxford University, obtaining an undergraduate degree in Philosophy, Politics and Economics and a postgraduate degree in Economics. After two years as a Research Fellow in Liecester University he moved to the Department of Social Policy in Edinburgh University, where he has lectured since 1971. He is the author of the following books — 'Investment Decisions in Nationalised Industries' (with R. Bates). Cambridge UP. 'Economic Policy Analysis : A Rights-based Approach' (with H. Neuburger). Avebury, and 'Childcare in a Modern Welfare System' (with B. Cohen), Institute for Public Policy Research, and a number of academic articles. His research has included work in health economics, on employment policies, and on social development

Anima Bhattacharya, MA (Calcutta University), MA (London School of Economics), retired from North Bengal University as a Professor of Geography in September 1997. She served in many educational institutions since 1954, including Jogamaya Devi College (then called Ashutosh College for Women) where she taught honours as a full-time Lecturer in Geography for three years up to 1960. On return from studying in London she was attached to a research wing of Calcutta University for a short time. She joined North Bengal University in 1965 as a Lecturer in Geography and then served the University continuously until her retirement. During this time she visited other universities like Burdwan, Calcutta, Visva Bharati, Gauhati etc. with academic assignments. She has published a number of papers and addressed various national and international seminars and conferences. She has written the following books, 'Population Geography of India', 'Samaj Bijnaniya Bhugol' ('Social Geography' in Bengali with B. Bhattacharya), 'Paschim Banga' (West Bengal), a study of its physical and social environment, in Bengali), and edited two books of papers from seminars she organised. She is still attached academically to Calcutta University, Gauhati University, and North Bengal University.

B. Bhattacharya, Educated at Calcutta University and (as Commonwealth Scholar) at the London School of Economics and Political Science, B. Bhattacharya has consistently pursued a career of teaching and research, having taught in many institutes, including Gauhati University (Assam), Indian Institute of Technology (Kharagpur, W. Bengal), Presidency College (Calcutta), and the University of North Bengal (W. Bengal), at the last of which he has been a senior Professor as well as the Dean of the Faculty of Science. His research in geography spans a large area with publications in academic journals and a number of books including 'Samaj Bijnaniya Bhugol' ('Social Geography' in Bengali) jointly with A. Bhattacharya, published by the West Bengal Govt., 'Urban Development in India since Pre-historic Times' (New Delhi), 'Sikkim — Land and People' (New Delhi).

GEOGRAPHY OF A HIMALAYAN KINGDOM : BHUTAN

NEIL FRASER
Senior Lecturer in Social Policy
Edinburgh University
United Kingdom.

ANIMA BHATTACHARYA
Retired Professor of Geography and
Applied Geography
North Bengal University, India.

BIMALENDU BHATTACHARYA
Retired Professor of Geography and
Applied Geography
North Bengal University, India.

CONCEPT PUBLISHING COMPANY, NEW DELHI-110059

ISBN: 978-81-7022-887-5

Edition 2019

Published and Printed by

Concept
Since 1974

Concept Publishing Company Pvt. Ltd.
A/15-16, Commercial Block, Mohan Garden
New Delhi-110059 (India)
T: 25351460, 25351794, *F*: 091-11-25357109
E: publishing@conceptpub.com, W: www.conceptpub.com
Editorial Office: H-13, Bali Nagar, New Delhi-110 015, India.

DEDICATED

TO

THE MEMORY OF

OUR PARENTS

FOREWORD

The Land of the Dragon

In the land of the dragon, the dzongs[1]
Have stood the Himalayan storms
Of elemental existence, carrying in
Their holy recesses, the silent norms
Of an old recluse's culture
Held within its silken folds
Of Buddhist texts, lit with
A hundred lamps, which hold
The key to the secret of its strength
In fortitude, a nation closed
Till of late, to the doors of the world
Waiting for the dragon who dozed,
To awake and let the shadows from across
The ranges roll over the Bumthang valley
Marvelling at the open countryside that was,
Till forty years ago, inaccessible, till
A frolicsome road looped round the bends,
Brazenly staring as Tangsibi[2] from the hill
Invading Ura La Pass, to ring the bell
With impish glee at its chorten[3]
Breaking the sanctity of Gankar Punsum[4]
— Devouring its virgin beauty with a wanton
Appetite, till the road comes to rest
In Ura village, creeping with docility through time
Into the new temple to Guru Rimpoche
Willing to wait, unwilling to startle the dragon in a fresh climb.

2 December, 1998 BASHABI FRASER

1. Buddhist monasteries
2. A village
3. A shrine holding a saint's relics
4. Bhutan's highest peak

PREFACE

Way back in mid-1979, a chance visit to Bhutan by one of the authors Anima Bhattacharya kindled the interest to write a book on this country which made a strong impression on her mind with its all pervasive Himalayan splendour. The idea gradually took a concrete shape during subsequent visits through years in which the other two authors Bimalendu Bhattacharya and Neil Fraser having developed a similar interest joined at a later stage.

In adding another to the list of books written on Bhutan so far, the interest was dominated largely by the idea of focusing on the present status of all the major aspects and bringing them within one frame.

Perhaps, the major hindrance we faced in writing this book was the lack of adequate data. Nevertheless, we are very much beholden to all the sources for their extreme kindness in helping us with data and information available with them.

We are thankful to Nirmal Chanda for cartographic assistance.

NEIL FRASER

ANIMA BHATTACHARYA

BIMALENDU BHATTACHARYA

CONTENTS

List of Plates

List of Figures

List of Tables

Glossary

Chiola	Offices
Chorten	Memorial Shrine
Chu/Chhu	River/Tower
Chunidom	Contribution/Tax (by labour)
Druk Gyalpo	His Majesty the King of Bhutan
Dungkhags	Sub-Divisional Headquarters
Dungtsho	Local Doctor
Dzong	Fort-Monastery
Dzongkhag	District
Dzongpon	Chief of the District
Gewog	Block (administrative)
Gompa	Small Temple
Laa	Mountain
Lakhang	Temple
Lam	Road
Lodoi Chodah	Royal Council
Lopen	Teacher
Penlop	Chief of the Provinces or Governor
Thanka	Scroll
Tshogdu	National Assembly

1

A BRIEF HISTORY

— Anima Bhattacharya

BHUTAN'S early history can be eked out from pre-historic Bhutan (600 B.C.—700 A.D.) when some other parts of the world witnessed the dawn of civilisation, now lost in oblivion without record. But Francois Pommaret believes that the country was inhabited probably around 2,000 B.C. This is ascertained by the stone implements found on the surface of the ground. This ancient country has meagre source of continuous narrative history. According to earliest legend, one Sangaldip from Cooch Bihar (erstwhile Assam) subdued Bengal, Bihar and spread his prowess as far as Bhutan in the seventh century before Christ. Both Indian and Bhutanese history agree to the fact that Bhutan was under Indian rulers for several centuries upto 650 A.D. The country was then ruled by Indian chiefs under the tutelage of Kamrupa. It then separated from Kamrupa when the latter disintegrated after the death of Bhaskaravarman. It was at this time that Tibetan incursion into Bhutan ensued.[1]

Bhutan's authentic chronicle unfortunately got lost in fire (printing establishment of Sonagachi in 1828, Punakha in 1832 and at Paro in 1896) and by the devastating earthquake of 1896, destroying the most valuable manuscripts of this Himalayan country. In such circumstances, one has to rely absolutely on history of Bhutan furnished by —

1. Accounts given by European travellers and from British Missions from 1774 to 1907 : George Bogle (1774); Alexander Hamilton (1775 & 1777); Samuel Turner (1783); Captain Pemberton (1838); Ashley Eden (1864); John Claude White (1905 &1907).
2. Accounts left by the East India Company and the British who built up relations with Bhutan from 1772 and signed regular treaties.
3. The few manuscripts found in the possession of the Tongsa Penlops which somehow survived the ravages of fire and earthquake. J.C. White, the British Political Agent in Sikkim, based his accounts on these manuscripts.
4. Some manuscripts available at Tibetan monasteries, found a place with Cheeboo Lama of Sikkim on which Ashley Eden had mainly relied.

Advent of Buddhism (7th and 8th century)

In the seventh century, the Tibetan King Srongtesen Gampo, a pagan by faith was converted to Buddhism by his own spouses, the first being a princess of the imperial house of China, the second a daughter of the Nepalese royal line, came to Bhutan in 640 A.D. and established two monasteries—Keychu Monastery at Paro and Jampa Lakhang at Bumthang. Thus, Buddhism was first introduced in Bhutan, besides Tibet and Nepal.[2]

Bhutanese tradition has it that Buddhism got widespread stronghold and deep rooted in the country with the appearance of Great Guru Padmasambhaba, who is also called Guru Rimpoche meaning precious teacher, in 747 A.D. Padmasambhaba came from Nalanda, a Buddhist monk who originally hailed from Uddiyana in N.W. India, to Bhutan at the call of King Naguchhi. King Naguchhi had founded the kingdom of Sindhu and his eldest son extended the realm as far as eastern Bhutan and in some parts of Tibet. Raja Nabudara who hailed from the plains of India defeated Sindhuraja Naguchhi whose eldest son was killed in the battlefield. It was at this time that the Indian saint appeared on the scene, saved Naguchhi, removed his grief for son's death and convinced him to Buddhist faith. The rival King Nabudara also accepted Buddhism and was converted to the faith. Peace was restored to the land and an era of Buddhism began in Bhutan to bestow the country with some sort of cultural unity. The Buddhism taught by the Guru is known as Ningma.[3]

Pre-Shabdrung Namgyal Period (9th — 15th and early 16th century)

Post-Guru period saw peace for nearly a century, when an apostate King of Tibet, Langdarma disturbed the stasis for a short period. Two centuries later, the Tibetans swarmed Bhutan to plunder, but ultimately liked the country so much that they did not want to return. They were called *Milog* (who did not return). From about the twelfth century A.D., many Lamas started pouring into Bhutan. It was at this time the dzong system of Tibet was applied to Bhutan. Lamas of the *Drukpa* (Red Hat) Sect came to Bhutan partly for missionary work and partly for the fear of persecution at the hands of the rival Yellow Hat Sect in Tibet called Gelukpa. The Drukpa, an off-shoot of the Nyingmapa Sect was founded at Ralung, about 48 km 30 Km east of Gyangtse in Tibet. Each religious reform in Tibet sent new stream of Lamas to the south, namely Bhutan. Earlier the Lama Head resisted the coming of the latter Lamas, but gradually such troubles came to an end in the long run either through the defeat of the attacking group or by acceptance of the new group by local villagers. At the end of the thirteenth century, a young Lama, assuming the name of Shigpo (Phajo-Drukgom-Shigpo from Sangye) built a small dzong named Dongon Dzong (Blue Stone Dzong) on the right side of the upper Wang Chu. With this, a new era started in Bhutan in establishing the church of Bhutan which persisted through centuries till today.[4]

Between the thirteenth and the sixteenth centuries, the Drukpa Sect (Red Hat) rose and consolidated defying all rival forces. This age saw great fervour of building monasteries and temples which gave Bhutan a distinct religious identity.

Rise of Shabdrung Ngawang Namgyal in the 17th century

Cultural unity under Buddhism could not stop internal fratricidal struggle in Bhutan among the multitude of Chiefs for holding their own territory.[5] Struggle for supremacy among the Chiefs gave rise to a man of destiny in the person of Shabdrung Ngawang Namgyal of a noble descent. He showed exceptional talent in his childhood. He was trained at Ralung and to avoid rivalry in the locality, he entered Bhutan by the Lingsi Pass at his twentythird year in 1616 A.D. He subdued Lhapha Kargyupa and other sects still extant there and established himself as the theocratic ruler of

Bhutan with the title Shabdrung Rimpoche on Dharmaraja. He had extraordinary power of organising and during his 35 years of rule, he consolidated his ability in spiritual and temporal authority. He met all opposition from Ralung, a brother of Dalai Lama V, and descendants of Lamas who came to Bhutan with success. He also repulsed the attack of Ghusuri Khan's Mongol-Tibetan troops in 1644. The defeated Mongol left large booty behind which swelled his coffers. His fame thus spread in India, Nepal and Ladakh. Shabdrung Namgyal received friendly missions and presents from these countries.

During Shabdrung's war with Tibet, two Portuguese missionaries, Estevao Cacela and Joas Cabral visited Bhutan. They met Ngawang Namgyal on way to Tibet and presented him some guns, gun powder and a telescope. They also offered their services against Tibet which was politely declined. Their visit is the first recorded visit of Europeans to Bhutan (Jesuit Missionaries, 1627).

The reign of Shabdrung Ngawang Namgyal witnessed the building of many monasteries and forts many of which were ravaged by fire and earthquakes and again were restored and enlarged. The Simtokha Dzong is perhaps the only structure which still stands as it was built in 1629. The next oldest dzong is that of Paro, originally built as a school of medicine and next rebuilt in 1646. Punakha Dzong built in 1637 was designed to accommodate 600 monks. The monastery of Wangdiphodrang had its origin in 1638, while the Tashi-Cho Dzong was built in 1641.[6]

The Age of Dharma Raja and Deba Raja

The great Shabdrung Ngawang Namgyal created and clearly defined the spiritual and political hierarchy of the country into the offices of Dharma Raja and Deba Raja (Desi). The bifurcated duties conferred the spiritual and religious affairs with the Dharma Raja (Shabdrung) and the other the Deba Raja with the general administration of the State. A Tibetan chronicler states :

> "In the interval of peace the Dharma Raja (Shabdrung) devoted himself with full energy to his various state duties, founding a body of priesthood, providing for and controlling them, giving instructions to those who were serious seekers after truth; in short, he was pastor, abbot, psalmist, rector, superintendent of carving

> (for printing purposes), architect of state and monastic buildings, overseer of book binding and other establishments of Kagyur library, settlement officer, chief commandant of the forces for quelling foreign aggressions, chief protector and ruler of his own adherents and followers, chief avenger and punisher of those who were inimical to the cause of Buddhism and the public peace. He was all these in one person, and fulfilled the duties thoroughly and efficiently. He introduced law in Bhutan...".[7]

The Dharma Raja thus was the epitome of law and justice and the Deba Raja wielded the function of a regent. Another distinction between the two institutions was that the Dharma Raja followed the incarnations of the predecessor, and the Deba Raja was elected by the council of permanent chosen members from the principal officers of the country. The Deba Rajas developed strength in course of time and became the temporal head of the State restoring material bounty.

The ecclesiastical authority of Dharma Raja continued its function as religious head at various levels. Monastery and dzong head Lamas were to follow the principles of righteousness, promote knowledge and protect religious sentiments. They should be properly trained for pious acts; gain necessary accomplishments in dancing, drawing or making mandalas besides acquiring knowledge of meditation. Those who are to acquire the other branches of learning, such as rhetoric, poetry, dialectics also must be encouraged.

The temporal authority of the Deba Raja performed the duties of administration and providing the Lamas with food. Deba Raja's authority was divided into the offices of Penlops (Chiefs of Provinces or Governors) and Dzongpons (Chiefs of Districts) for effective administration. Subordinate to the Dzongpons were the Neiboos, who supervised the scattered groups of villages.

Relation between India and Bhutan (1772 — 1865)

Sixteenth century onwards, with the establishment of Cooch Behar kingdom, it became the repeated target of Bhutan for incursions which increased both in intensity and frequency. The conflict continued till seventeenth century with the expansionist motive of both the States. The desire to spread their prowess beyond their kingdom's boundary was repulsed by retreat, promise of paying annual tribute or temporary treaty of maintaining no crossing of the proposed boundary line. At the same time Bhutan gave asylum to Cooch Behar Maharaja for safety when conquered by Mir Jumla. Bhutan also afforded protection to the infant Raja Debendra Narayan of Cooch Behar, who ultimately was murdered by Nazir Dev Rudra Narayan. Nazir Dev wanted to instal his nephew Khagendra Narayan as the ruler of Cooch Behar. Bhutan, however, was willing to put up the late Maharaja's step-brother on the throne, but the attempt was thwarted by Nazir Dev by seeking assistance of the East India Company.[8]

In the meantime, Bhutan kept Cooch Behar under control and carried out raids in opportune time. In 1771, they abducted the Raja Dhaijendra Narayan of Cooch Behar, the queen and the crown prince. At this crucial time, Nazir Dev of Cooch Behar approached the British for help which came very promptly.

The British intervention in 1772 opened a new era in the history of Bhutan , leading to its lasting relation with British India. On the 5th April, 1773, a treaty was signed by which Cooch Behar was required to pay a tribute of Rs. 50,000 immediately to meet the expenses of British troops. The British sent a small force to drive away the Bhutanese from Cooch Bihar and capture the three forts of Daling, Chitchacotta and Passakha. In desperation, Bhutan appealed to Tashi Lama, who was acting as the regent in Tibet during the minority of Dalai Lama and also his ally Prithvi Narayan Shah of Nepal. Tashi Lama's intercession led to a treaty of peace with Warren Hastings on the 25th April, 1774 at Fort William in Calcutta. The text of the treaty bound the Deba Raja to respect the territory of the East India Company, to refuse shelter to all who were hostile to the English, to permit the Company access to the forests under the hills and to protect the wood cutters. Bhutan conceded to deliver the captive Raja of Cooch Behar and pay a tribute of five Tangun horses for the fort of Chitchacotta. As the British had now access to Bhutan and Tibet, Warren Hastings decided to send trade-cum-intelligence mission to both the countries.

The first such mission was that of George Bogle of the Bengal Civil Service in May, 1774. The purpose of Bogle's Mission was to open a mutual and equal communication of trade between the inhabitants of Tibet and Bengal and to enquire into the nature of the road between them. This involved Deba Raja's consent to the passage of Indian traders to Tibet across Bhutan. Bogle was accompanied in his journey by Alexander

Hamilton as medical attendant. He took with him some presents like cloth of Britain and India, instruments , cutlery, hardware, firearms and valuables like strings of pearls, corals, brocades and shawls meant for articles of future trade between India and Tibet through Bhutan. He started from Cooch Behar via Buxa-Paro route of western Bhutan. On reaching Tashi-Cho Dzong, he was well received and had consultation with the Deba Raja about the prospect of trade with northern countries including Bhutan which lay on the transit. The dialogue yielded favourable result and the Deba Raja sought British assistance and protection for Bhutan's annual caravan to Rangpur.

Warren Hastings gave Bogle a free hand to make best possible use of his Mission to Bhutan and entrusted him with another Mission to Tibet in 1779 since Indo-Tibetan trade appeared as an assured possibility. Bogle also contemplated the establishment of extensive tea plantations in Bengal and Assam under British management. It was he who reported that in Tibet and Bhutan, tea was a universal beverage imported entirely from China.

Besides opening trade possibility in the highly inaccessible countries of Tibet and Bhutan, Bogle's Mission furnished great encouragement for the British to continue relationship with the Himalayan States.

Further, British Mission, the second in the series, was sent to Bhutan in 1775 under Hamilton , who earlier accompanied Bogle to examine the validity of Deba Raja's claims to the districts of Ambari-Falakata and Jalpesh. Hamilton came to a conclusion in favour of British interest.

The third British Mission to Bhutan again was sent under Hamilton in 1777. The invitation to Bogle to Peking in 1779 did not materialise due to the death of Bogle and the Lama who extended the invitation.

Warren Hastings next sent a Mission in 1783 under Captain Turner of the Bengal Army accompanied by Lieutenant Davis of the Bengal Engineers, as surveyor and Robert Saunders as the surgeon. The Mission travelling via Murshidabad, Rangpur and Cooch Behar was detained in Tashi-cho dzong for three months for want of permission from the Regent Teeshoo Lama to proceed further towards Tibet. The permission arrived to allow the Mission to move in two, leaving Davis to return to Bengal. Turner and Saunders proceeded further. There is no record on Turner's errand if it was charged with specific diplomatic assignment in Bhutan. However, the sixth mission of Eden to Bhutan states that Turner was authorised to cede to the Bhutan Government the district of Ambaree-Falakata, hitherto a British territory. The ceding was accomplished by Turner. He mentions in his report that the trade regulation of the Company through the dominions of Bhutan by the agencies of Indian merchants were settled by a treaty with Bogle in 1775. The treaty defined various rules applicable to the merchants of India and Europe and the specific articles reserved by the Deba Raja for trading between Bhutan and Bengal.

Comparative lull prevailed in the relation between Bhutan and the East India Company from the time of Turner's Mission 1783-84 to 1826 except for the visit of an Indian officer Kishen Kant Bose sent by the Judge of Rangpur seeking settlement for incessant border disputes. Kishen Kant reached Punakha via Goalpara, Bijni, Sidlee, Chirang, and the valley of Pho Chu and Mo Chu rivers. His report was comprehensive and intelligent, assiduously collected from local sources and personal observations. Kishen Kant found that the country was then ruled by a Raja of Cooch tribe. Lam Zhapte wielded miraculous powers, and drove out the Raja and his tribe by playing on a pipe made of human thigh-bone and by acting contrary to the observations of Cooch tribe. Thus, he got the possession of Punakha and the country of Bhutan. After occupying entirely the religious domain, he appointed a Tibetan from Lhasa as the Deba to look after the affairs of the government. The dual system thus established lasted for many centuries in Bhutan. The account of Kishen Kant Bose, however, does not conform to the established historical facts.

Trouble between Bhutan and the British concerning Duars

At the base of the lower ranges of Bhutan hills, there is a narrow corridor of rich and fertile tract, ten to twenty miles in width, extending from Dhansiri in Assam in the east to Tista in the west in Darjeeling. This tract holds as many as eighteen duars, seven in Assam and eleven in Bengal. These are known as Bhutan Duars (passes). Bengal's eleven Duars are situated between the Tista and the Manas rivers and Assam Duars between the Manas and the Dhansiri rivers. These duars were inhabited by the Mechis and the Kachari tribes and were administered by the Bhutanese Jongpens. Local administration of these duars was in the hands of the Bengalis, the Assamese and the Kacharis who were appointed by the sanad of Deba Raja. The control of the Bengal duars was wrested by Bhutan from the Muslim ruler, but the Assam duars were never under its control.The Ahoms could not maintain law and order there and the security of this tract was

Plate 1 : A Chorten

Plate 2 : One of the earliest Buddhist Monasteries, Paro

exchanged between two countries by an annual payment of yak tails, ponies, musk, gold dust and blankets. These arrangements were so complicated that interminable disputes became inevitable when the British occupied lower Assam. As Bhutanese wares were in kind, their actual value often fell below Bhutanese assessment which led to payment in arrears by Bhutan Government. Whenever clearance of arrears were sought, Bhutan retaliated by plundering the border areas (the duars).

After the first Burmese War (1824-26) when the British had driven the Burmese out of Bhutan, the extent of Indo-Bhutan boundary increased and with it increased the disputes between the Assamese and the Bhutanese. The first attack on Chatgar in Booree Goomah Duar in 1828 embittered the relations between the British Government and the Bhutanese ruler which continued till 1864. In 1836, the British attached Baksa Duar for arrears of payment due to British. The Deba Raja and Tongsa Penlop pleaded that the communication of arrear payment to the Bhutan Government did not reach and the payment may be realised from the Baksa Duar and the Duar itself if the areas are restored to Bhutan. The British agreed to surrender the Duar on the condition that Bhutan will manage them properly in future. According to Eden, the agreement was unfortunately signed by a Bhutanese official of lower rank and was never ratified by the Deba Raja.

In 1838, the Governor General sent an envoy of Captain Pemberton to Bhutan to settle the commercial intercourse between the two countries. The unfriendly relation led to the failure of Pemberton Mission and the draft treaty proposed by him was rejected by Bhutan rulers. It was the Tongsa Penlop who objected to the treaty as his interests were directly related to the punctual payment of the tribute for the Assam Duars.

The failure of Pemberton Mission led the British to think about permanent annexation of the Duars. In 1841, all the Assam Duars were taken over in return for an annual compensation of Rs.10,000. In 1842, the estate of Ambaree Falakata was also taken over at the request of Bhutan. This annexation fulfilled a major motive for economic gain of the British linked with the growth of tea plantations in the area.

The annexation, however, did not solve the problems in the duars which continued for years and Bengal Duars particularly remained a trouble spot. The annexation of the Bengal Duars was deferred by serious problems of the British elsewhere concerning the Afghan war and the Anglo-Sikh wars in Sind and the Punjab. It was not until 1850 that Lord Dalhousie started action by appointing Jenkins to be in-charge of the Bhutan frontier. The Bhutanese raids on Goalpara, Cooch Behar and Rangpur frontiers were resisted by the British arms. As a sequel to this, the British Mission under Ashley Eden was led to Bhutan in 1864.

Ashley Eden of the Bengal Civil Service was chosen as the Envoy for a special mission to Bhutan. Captain Godwin Austen was appointed as Assistant to the Envoy. Dr. Simpson as Medical Attendant, the Chebu Lama of Sikkim as Tibetan Interpreter, and Captain Lance of the Bengal Staff Corps was to command an extraordinarily large escort of 100 troops, half Sikhs and half Sappers of the Assam Sebundy Corps.

The Eden Mission arrived in Darjeeling in November 1863 where it was detained for more than a month for the Bhutanese reluctance to receive the Mission for its own internal civil war. Eden's fateful Mission started in January, 1864 and reached Punakha on March 15, 1864. He found the new Deba Raja was a stooge in the hands of Tongsa Penlop, who never forgot his misfortune with the loss of Assam Duars at the hands of the British. The ill-omened British Mission is generally regarded as a mission of forced intimacy.

The Bhutan Government was disinclined to receive Eden's Mission. Ignoring this Eden continued his journey to Punakha. At Simtokha, he found the dethroned Deba resting. The ex-Deba declined to meet the British Mission that held communication with the 'Ferengs'.

The Tongsa Penlop demanded that the affairs of Assam Duars should resume in the process of discussion and to this Eden replied the matter had been closed so he had no authority to raise it further. This infuriated the Tongsa Penlop who declined to treat the Mission any further if the British usurpation of the Assam Duars were not settled. Wangdiphodrang Dzongpon also charged the Chebu Lama for bringing in the British Mission.

In the session with the Council, over which the Tongsa Penlop presided, the draft treaty was read article by article. At the meeting, the Tongsa Penlop proposed that Assam Duars should be restored to Bhutan and the revenues collected from them since the date of resumption amounting to Rs.300,000 per year should also be paid to Bhutan. Eden emphatically replied that he had no authority on the subject and summarily rejected the proposal to discuss. Heated arguments followed with tempers and both sides became offensive. At one stage, infuriated Tongsa Penlop picked up a large piece of wet dough, rubbed that on Eden's face, pulled his hair, slapped him on the back to humiliate Eden utterly. Wangdiphodrang

Dzongpon went even further who in a gesture of familiarity took out a bit of chewed *pan,* (betel leaf) from his mouth and asked Dr. Simpson to eat that. When Dr. Simpson refused, he angrily spitted on his face. The Chebu Lama's watch ribbon was taken from his neck and was wrenched away. Tongsa Penlop declared at this point that he wanted nothing but the Assam Duars, and if he did not get that, it was better to have war than a treaty. He further stated that he was going to write to the Governor General.

The British Envoy was subjected to sign a treaty prepared by the Penlop after humiliating treatment. The Tongsa Penlop and Deba's council highly alarmed, advised Eden to withdraw his Envoy at once from Punakha. Eden could not abide by the counsel. The Tongsa Penlop doubted Eden's authority and made him to leave Punakha under compulsion. The Mission then took a hurried departure, after taking leave from the Deba and the spiritual ruler with a letter for the Governor General along with three ponies and a few pieces of silk as presents.[9]

The treaty which Eden had to sign under duress stipulated that there shall be eternal peace and friendship between the Ferenga and the Bhutanese. Bhutan held responsible evil men on both sides who created rift between the British and the Bhutanese Government. The British would surrender all Bhutanese territories at the frontier and thereafter, each will take charge of its own territory without committing any aggression and thus maintain friendly terms. No mention was made about the compensation for the Assam Duars. Whichever of the four States — Bhutan, the Ferengs, Cooch Behar and Sikkim—resorts to aggression, the other three shall punish it.

The Anglo-Bhutanese War (1864-1865)

In utter disgrace, the crestfallen British Envoy left Bhutan and returned to Darjeeling by the end of April, 1864. At that time both Shabdrung and the Deba were helpless puppets in the hands of notoriously unscrupulous robber Chiefs — the Tongsa Penlop and the Wangdiphodrang Dzongpon. On May 7, 1864, Eden communicated to Fort William suggesting adoption of one of the three measures against Bhutan to secure British frontiers : (i) permanent occupation of the whole country, or (ii) temporary occupation of the country and destruction of all forts in Bhutan or, (iii) permanent annexation of the eleven Bengal Duars.

Eden's Mission was subject to acrimonious comments. He was publicly censured for his unwise insistence on proceeding to accomplish the objectives of the Mission which the Bhutan Government emphatically declined to receive.

The Government of India first announced permanent annexation of Ambaree-Falakata and withheld the annual compensation for the Assam Duars. At the same time, repatriation of all British and Cooch Behar subjects in captivity were demanded upon the Bhutan Government. Unless the demand was fulfilled, permanent annexation of the Bengal Duars was threatened.

The British demand was rejected by Bhutan and a 10,000 strong British force was assembled on the Indian frontiers to advance into the Bhutanese territory from different points.

When military preparation for the annexation of the Bengal Duars and the capture of all the forts in Bhutan was complete, the Governor General of India on November 12, 1864, issued a proclamation of war against Bhutan. The latter was accused of committing outrages within the British territory, the territories of Sikkim and Cooch Behar. Bhutan was charged for insulting the British Envoy in open court and compelling him to sign a treaty which the British repudiated. The Bhutan Government strongly rejected the British accusations. The rejection of the other demands of the British Government by the Bhutan Government led the Governor General to occupy the Bengal Duars and the hill territory of Bhutan including the forts of Dalimkot, Pasakha and Diwangiri.To give effect to the measures, the British force advanced towards Bhutan from Gauhati (to Diwangiri), Goalpara (against Bishen Singh), Cooch Behar (against Buxa and Bala), from Jalpaiguri (against Dalimkot and Chamurchi).[10]

The Bhutan Government shocked at the British gesture of violating the peace treaty made with Eden, the Deba contended that if the British possessing a large territory was inclined to take possession of the small territory without fighting, he would revenge by sending divine force of twelve gods, who were very ferocious ghosts. Of this force 7,000 stopped at Chamurchi, 5,000 at Durma, 9,000 at Buxa, and 102,000 at Dalim Duar.

The threat given by the Deba Raja of invoking the ire of twelve gods did not prove to be unfounded. The Tongsa Penlop, soon after, swept down upon the eastern post of Diwangiri with a strong Bhutanese force and cutting off its water supply, compelled the evacuation of the British on February 5, 1865. The complete disorganisation in the British force was experienced following similar attack all along the occupied territory. General Malcaster and General Dunsford were replaced by

Brigadier-General Tombs C.B., V.C. and Brigadier-General Fraser Tytler, C.B. respectively. It was not until March 1865 that the British could make a counter attack and recapture the posts evacuated earlier. A complete lull followed thereafter throughout the summer of 1865 and this led to an environment of negotiation for settlement. The British threat of further military action in extensive scale made the Bhutanese relent in October, 1865 and the Deba Raja offered to negotiate with the British through his officers.

The peace negotiations resulted in the famous Sinchula Treaty in November, 1865. This treaty is known to the Bhutanese as the Ten Article Treaty of Rawa Pəni and will remain as one of the landmarks in the annals of British-Bhutanese relations. The main issues of the articles were :

Article 1	Perpetual peace and friendship between the two Governments.
Article 2	In consequence of repeated aggressions of the Bhutan Government, the British concluded that the eighteen Duars together with the Talook of Ambari-Falakata and the hill territory on the left bank of the Tista were to be ceded by the Bhutan Government forever.
Article 3	The Bhutan Government surrendered all British subjects as well as the subjects of the Chiefs of Sikkim and Cooch Behar who were detained against their will. No impediments were placed on the way of return of such persons into the British territory.
Article 4	Bhutan expressed its regret for past misconduct and agreed to receive an annual allowance from the British Government not exceeding fifty thousand rupees.
Article 5	The British Government might suspend the payment of the compensation money either in whole or in part in the event of misconduct or its failure to check the aggression of its subjects.
Article 6	The British Government hereby agreed to surrender all the Bhutanese subjects accused of any of the following crimes— murder, attempting to murder, rape, kidnapping, great personal violence, maiming, dacoity, thugee, robbery, burglary, knowingly receiving property obtained by dacoity, cattle stealing, breaking and

entering a dwelling house and stealing therein, arson, setting fire to village, house, or town, forgery, or using forged documents, counterfeiting current coin, perjury, subordination of perjury, embezzlement by public officers or other persons, and being an accessory to any of the above offences.

Article 7 The guilty persons committing above crimes would be surrendered by respective governments on requisition being duly made by the authority of the local court of the district or by the authority of the Lieutenant Governor of Bengal.

Article 8 The British Government hereby engaged to enquire into and settle all disputes and complaints in such manner as justice may require and to insist on the observance of the British decision by the Rajas of Sikkim, Cooch Behar and Bhutan. This gives a position of the political superiority to the British. The British Government gets installed as the Supreme Court of appeal over disputes of the Rajas.

Article 9 There shall be free trade and commerce between the two Governments and no duty should be levied on the goods imported by either side. Subjects of both the territories will get equal justice.

Article 10 The treaty of Sinchula thus concluded, signed and sealed by the authority of Lieutenant-Colonel Herbert Bruce C.B. and Samdojey Deb Jempey and Thimseyrensey Donai, the ratification of the same by His Excellency the Viceroy and Governor General and by their Highnesses the Dharma and Deba Rajas shall be mutually delivered within thirty days from this date.

From 1865 to 1910

Political situation on the turbulent frontiers quietened after the Sinchula Treaty. In the period of comparative peace, internecine rivalries which have been the bane of Bhutan, came to the surface in 1869. A civil war broke out in Bhutan in 1869 in which Tongsa and Paro Penlops along with the Punakha Jongpen rebelled against the Deba Raja who was supported by the Wangdiphodrang Jongpen. Being approached by both the parties, the British refused to intervene in the internal affairs of the State.

In 1877, another civil war broke out by the pretensions of a rival claimant and Jigmey Namgyel assumed supreme authority to crush the rebels. Another contest for the office of the Deba Raja occurred in 1880. In course of intermittent civil wars, the civil war of 1885 proved decisive and Tongsa Penlop, a strong man, emerged as the virtual ruler of Bhutan. To consolidate his position, the Tongsa Penlop sought the friendship of the British who readily offered their support for this strong man who could guide the destinies of Bhutan. In 1899, the British sent an expedition to Sikkim. The Bhutanese could not be persuaded by the Tibetans into joining them against the British.[11]

Both the Paro and Tongsa Penlops visited the British Political Agent at Kalimpong to obtain the British friendship. The British Mission to Tibet under Younghusband was taking place at that time and Ugyen Wangchuk was instrumental in assisting Colonel Younghusband to contact with Tibetan officials during the march to Lhasa. When the British Mission made way into Tibet in 1904, the Tongsa Penlop showed his keen interest in the negotiations to have closer ties with India. The Indian Government was equally keen on having ties with Bhutan through the lines of communication from India to the trading posts in Tibet. From 1903, the political affairs of Bhutan were no longer looked after by the Government of Bengal. The British Mission to Lhasa was also authorised to correspond directly with Bhutan. Indo-Bhutan friendship was further strengthened in 1905, when a Political Agent was appointed by the centre in direct relationship with Bhutan. This relationship reduced the mutual suspicion between India and Bhutan.[12&13]

In 1905, the political officer J.C.White presented to the Tongsa Penlop the insignia of the order of the Knight Commander of the Indian Empire. As against the ill-omened Mission of Eden 40 years ago, the White Mission of 1905 was received with all warmth, enthusiasm and friendliness of which White spoke so highly in his publication — 'Sikkim and Bhutan'.[14]

The Tongsa Penlop had greatly strengthened his position by 1907 and was progressing fast to assume the headship of the State. It was the eventful year which witnessed the unique election of Ugyen Wangchuk by a unanimous vote of Bhutan Chiefs and the principal Lamas as the hereditary Maharaja of Bhutan. The British and the Political Agent John Claude White helped in the establishment of the hereditary monarchy.

The installation ceremony took place on December 17, 1907.

The creation of hereditary monarchy led to internal political stability and peace and cordiality between British India and Bhutan.

The far-sighted Ugyen Wangchuk was quite aware of the fact that Bhutan needed concrete steps to ensure material well-being of the State. His first step was, therefore, to open Bhutan to Indian agencies to develop its resources. Various schemes and projects covered a large area like schools and education, population, trade, the construction of roads, the exploitation of the mineral resources of the country and the best method of utilising them besides encouraging tea cultivation on the wastelands at the foot of the hills.

The Period between 1910 and 1947

The Treaty of Punakha signed by His Highness Sir Ugyen Wangchuk, the Maharaja of Bhutan and Mr. C.A. Bell, Political Officer in Sikkim on January 8, 1910 was subsequently ratified by Earl of Minto, the then Viceroy and Governor General of India. This treaty amended Articles 4 and 8 of the earlier treaty concluded at Sinchula on November 11, 1865. Article 4 recommended an increase of the annual allowance from fifty thousand to one hundred thousand rupees. Again, Article 8 of Sinchula Treaty was revised to make British non-interference in Bhutan's internal affairs dependent on Bhutan's acceptance of the British advice in Bhutan's external affairs.

Again, another agreement was signed in 1914-15, for capturing elephants in the districts of Assam and the contiguous areas of Bhutan to improve the finances of the State. While elephants abounded on Indian side, the best stockade sites lay in Bhutan. The area in question was to be mapped out into three *mahals* — named Goalpara, Kamrup and Darrang *mahals* — during 1915-17, 1917-19 and 1919-21 respectively. The profits derived from the operation were to be shared equally by Bhutan and Assam. Bhutan, however, reserved the right of closer control over grazing, poaching and firing by the Nepalese settlers in Bhutan, in the adjoining reserve forests of Assam.

King Ugyen Wangchuk, Bhutan's man of destiny died on August 21, 1926 after a successful reign of 19 years. He was succeeded by his son Druk Gyalpo Jigme Wangchuk in 1928. The British Empire reached the pinnacle of its glory both at home and abroad with the silver jubilee celebrations of King George V in 1935 which kept continued friendly terms with the inaccessible mountain kingdom of Bhutan.

Advent of Modern Bhutan since 1947

Before 1947, Bhutan never acknowledged suzerainty of any external power. With the departure of the British, Bhutan took a step towards its emergence as a modern State. Bhutan'a relations with free India, its nearest neighbour — whose foreign policy was committed to promote the decolonisation of countries in the continents of Asia and Africa, whole-heartedly assisted Bhutan to rise to her full stature in the following years. India pronounced that Bhutan was no part of the territory of India and was, therefore, a totally autonomous, independent political unit whose external relations could only be governed by the consent of Bhutan.

The Indo-Bhutan Treaty of 1949 contained ten articles, the cardinal texts of which were : there shall be perpetual peace and friendship between the two Governments; India will not interfere in the internal affairs of Bhutan and Bhutan agrees to be guided by the advice of the Government of India in regard to its external relations; compensation granted to the Government of Bhutan to be enhanced and as regards the temporary subsidy of Rupees one lakh per annum granted in 1942, India agrees to make an annual payment of Rupees five lakhs to the Government of Bhutan; there shall be free trade and commerce between the territories of India and Bhutan.

Among the remarkable changes in Bhutan's political affairs recorded since 1947, is the country's membership of International Community with its admission to the membership of the Colombo Plan in 1963.The country's second historical landmark was its admission to the Universal Postal Union in 1969. Thirdly, Bhutan's admission to the United Nations may be regarded as the culminating point of development which was initiated by India on December 10, 1970. The King as the Head of the State sent an application for admission to the United Nations. The Committee on admission of the Security Council recommended the admission of Bhutan on February 9, 1971 and the Council unanimously adopted it. On February 10, 1971, India applauded Bhutan's candidature for admission to the United Nations. In this way Bhutan was recognised by the nations as an independent State in the international community since February, 1971. Same year it became a member of the G-77 Group and has been a member of the SAARC since its inception in 1985.

Changes in Administration

In the sphere of administration, the country has experienced remarkable changes since ancient times. In the past, the country was divided into a number of social units administered by local chieftains. As observed earlier, the concept of kingship, however, appeared as cardinal political institution based on the principles of Aryavarta in its hereditary form, supreme in all affairs of the life of a region. Medieval Bhutan records dual authority, one supreme in temporal affairs and the other in spiritual affairs as was known in Medieval Europe. During this time religion played a great role in the politics of the country and was greatly influenced by the Tibetan Order in the cultural life. Monarchy, however, in Bhutan has undergone vicissitudes of historical changes in conception from single hereditary institution in the past to an elective position of headship in the middle ages in temporal affairs with an equal status to the spiritual Head, the process which lasted in Bhutan till the nineteenth century. During these times, the country was constantly engaged in internecine strife and feudal anarchy of the worst order. This state of affairs led to the necessity of the establishment of a unified central authority to save the country from utter disintegration.

Shabdrung Namgyal, the leading incarnation following Pamkarpo entered Bhutan through the Lingzhi Pass in 1616 A.D. Shabdrung founded the independent theocracy of the 'Land of the Thunder Dragon', Drukyul. He represented in personal union, the spiritual and the temporal power of the State, which, by his incarnated followers, was divided into the position of a Dharma Raja and Deba Raja. The 'Iron Monastery', Cari Gompha in the north of Thimphu, is a memorial to his first resting place in western Bhutan. Its temple hall is graced by series of painted scrolls depicting the line of eleven Shabdrungs whose incarnation lasted for nearly 300 years and ceased to take place only after 1905, a date which marks the beginning of the new era of Bhutan. During all these times Bhutan was torn with intermittent strifes and the country's integration was strongly felt. It was rather an historical event in which the effort of Tongsa Penlop Ugyen Wangchuk built himself to the supreme status among the other Penlops and heads to salvage Bhutan from its chaotic condition to a cosmic position. The elective method of Headship which emerged in the middle ages, in rudimentary form, became effective vividly in the election of the Tongsa Penlop as the Head of unified Bhutan under the positive support of the

British diplomacy. Thus, on December 10, 1907, the Tongsa Penlop, Ugyen Wangchuk, the Prime Minister of Bhutan was installed as the hereditary Maharaja of Bhutan in the presence of the Abbots, Lopens, the whole body of the Lamas, the State Councillors, the Chiolas of different districts and the distinguished subjects. So with the end of Namgyal rule in 1905, a new era began in the country in 1907 when the present hereditary royal line was established.

Social changes in administration since 1907 is characterised by democratic principles where administrative functions are discharged by a National Assembly (Tshogdu), an advisory council and a regular Central Secretariat. In short, the machinery of the Government bases on four broad tiers namely : (a) the central machinery of the executive and the legislative at the capital; (b) the local administration at the villages and the districts; (c) the Judiciary; and (d) the armed forces. Druk Gyalpo, the King of Bhutan in this set-up, was not only the Head of the State, but also the Commander-in-Chief of the armed forces, the highest court of appeal and the wielder of commanding influence in religious and spiritual matters.

The office of the Dharma Raja was thus totally abolished under the control of the King.

In addition to the National Council, the king established in 1965 the Royal Council (Lodoi Chodah) which is in session throughout the year as against Tshogdu which meets twice a year. The Council consists of a Chairman, people's representative of the Government, enjoying the rank of a minister called Kalan, and monastic representatives.

The constitutional reforms by King Jigme Dorji Wangchuk by establishing the Tshogdu or the National Assembly of Bhutan, which transformed the country into constitutional monarchy in 1953, initiated in Bhutan a system of government based on democratic principles. It included the elected representatives of the people, of the monastic order and official nominees. The people's representatives were to be elected by indirect method based on population; the monastic representatives were nominated by the various monk-bodies; and the official representatives were nominated by the King from among the senior civil and judicial servants on functional basis. The National Assembly of Bhutan became the chief legislative organ of the State by judicial reforms.

The members of the Tshogdu who have freedom of speech, hold office for three years and meet twice a year. The powers and functions of the National Assembly are to enact laws, approve all senior appointments made by the Executive, and discuss and advise on all matters of national importance.

The King usually nominates his Council of Ministers, but the appointments of the ministers must be approved by the Tshogdu, which has also the right to nominate ministers and remove them by a majority vote any time. The salaries of the ministers are voted by the Tshogdu. All ministers are of equal rank.

REFERENCES

1. Aris, Michael 1980, *The Early History of a Himalayan Kingdom*, Vikas Publishing House Pvt. Ltd., Delhi.
2. Deb, Arabinda, *Bhutan — A Study in Frontier Political Relations* (1772-1865), Firma K L Mukhopadhyay Ltd., Calcutta.
3. Edmunds, T.O. 1988, *Bhutan—Land of Thunder Dragon*, Elm Tree Books, London.
4. Hasrat, J.H. 1980, *History of Bhutan — Land of the Peaceful Dragon*, Education Department, Royal Govt.of Bhutan.
5. Karan, P.P. 1967, *Bhutan—A Physical and Cultural Geography*, University of Kentucky Press, Lexington, U.S.A.
6. Mehra, G.N., 1978, *Bhutan—Land of Peaceful Dragon*, Vikas Publishing House Pvt. Ltd., Delhi.
7. Noble, I., 1984, *Bhutan — A Kingdom of the Eastern Himalaya*, Geneva.
8. Olschak, B.C., 1979, *Ancient Bhutan—A Study on Early Buddhism in the Himalaya*, Swiss Foundation for Alpine Research, Zurich, First Edition.
9. Pommaret, F., 1991, 1990, *Bhutan*, (Translated by Elisabeth B. Booz), Geneva.
10. Ronaldsay, 1977, *Himalayan Bhutan, Sikkim and Tibet*, Ess Ess Publications, Delhi.
11. Rustomji, N., 1978, *Bhutan — The Dragon Kingdom in Crisis*, Oxford University Press.
12. Singh, N., 1978, *Bhutan—A Kingdom in the Himalayas*, Thompson Press (India) Ltd., New Delhi.
13. Turner, S., 1800, *A Journey Through Bhutan*, London.
14. White, J.C., 1971, *Sikkim and Bhutan*, Vivek Publishing House, Delhi.

2

PHYSICAL BACKGROUND

— Bimalendu Bhattacharya

BHUTAN, known to people as Druk Yul or Druk, the Land of the Thunder Dragon, is situated on the southern slopes of the eastern Himalaya and lying between 26.5 and 29 degrees north latitudes and 88.5 and 90 degrees east longitudes, it comprises an area of 46,590 sq km. It is of compact shape, stretching for about 150 km from north to south, and about 300 km from east to west but due to mountainous terrain, the actual distance on the ground is more than double. Surrounded by high mountain ranges, it is a completely landlocked country with a very rugged terrain, bordered to the north by the Tibetan region of China and makes frontiers with the Indian States of Sikkim, West Bengal, Assam and Arunachal Pradesh in the west, east and south. In the north-west, the stark mountains of the Chomolhari Range provide the frontier with the Chumbi Valley of Tibet and in the far north lies the snow ranges of the Great Himalaya making a formidable barrier between Bhutan and China. It is in the extreme south, however, that the ruggedness disappears largely in the foothill region of the Southern Duars bordering the Indian States of Sikkim and West Bengal where the average elevation of the land comes down to about 160 m above sea level from more than 7,000 m in the north, undoubtedly a great variation over a distance of barely 150 km. There are six major mountain ranges in Bhutan running from north to south more or less parallel to each other, giving rise to seven drainage systems in between them with southward flow. From west to east they are — Ammo Chu or Torsa, Wang Chu or Raidak, Mo Chu or Sankosh, Mangdi Chu, Bumthang Chu, Kuru Chu, and Kulong Chu which join the river Brahmaputra in Assam.

Geologically, the area occupied by Bhutan is considered to represent the old Tethys Sea of the geological era later on transformed into the confused systems of mountain ranges at the time of great mountain building during the Tertiary period. Hence, the mountains are of recent origin consisting largely of pre-Cambrian and early Palaeozoic quartzites and gneiss. Some sedimentary areas of limestone, dolomite, sandstone and shale are also found, especially in the Manas basin. The massive ranges were, naturally, the products of erosive processes continuing for millions of years giving rise to extremely rugged topography represented by lofty mountain ranges interspersed by ravines.

Bhutan has drawn the interest of many geologists , noted among them being Austin (1868), Mallet (1875), Pilgrim (1906), Halden (1907), Lahiri (1941) and Gansser (1964) who have worked on and contributed profusely to unravelling its highly complex geology which underwent intense tectonic activities involving intricate folding, faulting and thrusting processes in the geological past. However, the absence of fossils in many of the rocks makes the study of the geological structure quite difficult.

The various studies made by different geologists reveal that "some of the sedimentary formations have been vertically lifted about 4 to 5 km above their original level of deposition and some of the rock units have also undergone large scale horizontal trasportation from north to south over many kilometres"[1], giving rise to complicated geological formations.

Traversing from south to north one comes across the following rock formations :

(a) The Siwalik (Miocene to Pleistocene), composed of sedimentary rocks, the bedded sandstones, claystones and occasional nodules of limestone being the most dominant types, occur in the foothill region east of river Raidak and in eastern Bhutan near Samdrupjongkhars;

(b) The Gondwana (Permo-Triassic) rocks consisting of quartzitic sandstones, carbonaceous shales and some coal seams with abundant fossils lying over the younger Siwaliks are exposed in the foothill region between Samdrupjongkhar and Daifam;

(c) The Buxa Formation (Lower to Middle Palaeozoic) comprising mainly weak metamorphosed sedimentaries represented by thick green and purple dolomite bands, carbonaceous phyllites and quartzites occur as discontinuous patches all through the foothill region;

(d) The Duriri Formation (Carboniferous) with pebbly phyllites, shales and quartzite as the dominant rock types exposed in the eastern Bhutan forms a part of the Buxa Formation;

(e) The Daling Formation (Lower to Middle Palaeozoic) characterised by a green phyllite with a soapy feel along with the occasional presence of grey phyllite, quartzites and thin bands of crystalline limestone occurs as a continuous exposure in the foothill and the lower Himalayan zone of the western and central Bhutan;

(f) The Shumar Formation (Lower to Middle Palaeozoic) comprising alternating sequence of green and grey phyllites with limestone and quartzite occurs in the lower Himalayan zone of the eastern Bhutan;

(g) The Thimphu Gneissic Complex (Lower to Middle Palaeozoic) with gneiss, high grade schists and bands of quartzite and marble as the major rock types, lies over the Daling, Buxa and Shumar Formations are found in the central Bhutan;

(h) The Paro Formation (Lower to Middle Palaeozoic) consisting of quartz-mica schists, quartzites, calc-silicate rocks, marble and thin bands of graphaitic schists occurs within the Thimphu Unit.

(i) Chekha Group (Lower to Middle Palaeozoic) comprising flaggy quartzites, crystalline limestone, green and grey quartzites and phyllites etc. covers a large area in the western and the central Bhutan.

Besides, the Tonghu Series/Crinoidal Limestone (Devonian) mainly composed of limestone and slates, the Shodug Series (Upper Carboniferous) consisting of pebbly slates and quartzites, the Upper Gondwana (Jurassic) made up of quartzitic sandstones, shales, carbonaceous shales etc., the Nigle-La/Lingshi Series (Jurassic to Cretaceous) represented by dark grey slaty shales and the Chomo-Lahri granite (Tertiary) occurring in the west of Tongsa intrusive in character, are significant.[2]

Physiographically, Bhutan may be divided into three distinct units : the Great Himalayan Zone of the north and north-west; the Inner Himalayan Mountain and Valley Zone representing the central part of the country, and the Sub-Himalayan Foothills Zone of the south known as the Lower Himalaya.

The Great Himalayan Mountain rises in the extreme north, dividing

Bhutan from the Tibetan Plateau and consisting of giant snow peaks it makes a very bleak, barren landscape with an extremely rugged topography. Across the region stretches Lunana, an area of sharp etched glaciers and exposed rock faces with snow-covered mountains and deep glacial lakes, making the watershed of the Pho Chu, one of the wildest of the country's rivers. Thanza, the highest settlement of Bhutan lying at an altitude of about 3,750 m in eastern Lunana, is approachable only through steep passes. Lunana, Lingshi and Laya may be considered as high valleys inhabited only by yak herders where cultivation is limited by the high elevation. The region is composed of sedimentary rocks ranging in age from Cambrian to Tertiary. Some of the loftiest, perpetually snow-clad peaks occur in the north and north-west. They are Thongsey Devi Gang in central Lunana; Jejekangphu Kang (7,315 m) in the north-west; Gangker Punsam (over 7,498 m) in the east; Jomo Lhari (nearly 7,315 m) in the north west; Kulha Gangri (7,540 m) in the east. The rivers rising from the high mountains and flowing southward have carved out a complicated labrynth of deep-set narrow valleys at the foot of the glaciers at an elevation of 3,660 to 5,490 m.

There are four passes in this region, three in the west and one in the east through which Bhutan has maintained trade and cultural relations with Tibet over centuries till recent times, keeping contacts between Lhasa and Paro, Punakha and Tashigang trading in rice, fine silk fabrics, dyes and brass utensils as export items and salt, soda, wool, gold dust and silver as imports. With the cessation of this trade from around 1959, when Tibet was invaded by China, the trade-posts lying on these routes were abandoned, taking today a forlorn look.

The Inner Himalayan Zone, comprising the central part of Bhutan, consists of mountain ranges as high as 3,660 to 4,570 m, and running from north to south form major watersheds between the principal river systems. The Black Mountain Range forming the major divide, and separating the waters of the Sankosh and Manas river systems makes also a cultural divide between the east and west of the country. For instance, in the west the people are mainly of Tibetan origin while in the east live the descendants of the original inhabitants of Bhutan, a people whose ancestry has yet to be conclusively traced. Pele La, a 3,300 m high mountain pass, makes the contact between the east and the west of Bhutan. The rivers here are turbulent, rushing through gorges before they empty on the Indian plains to become large tributaries — namely Torsa, Raidak, Sankosh and Manas — of the Brahmaputra river.

The most conspicuous relief of the Inner Himalayan Zone is represented by the wide to moderately wide river valleys providing undulating land at high altitudes carved out of a mountainous terrain. In the east the land consists of three sub-regions of Mongar, Tashigang, and the area extending upto Samdrupjongkhar on the Indian border. It is a land of deep V-shaped valleys with fields and dwellings perched on the bare slopes. At three days walk from Tashigang eastward, lie the high valleys of Merak and Sakteng inhabited by herdsmen. Each of these valleys separated from one another by lofty mountains often 4,000 — 5,000 m in elevation has developed as individual units connected with the next valley by a high pass (average altitude 3,000 m). Communication between them is thus quite difficult though a major road now links up all the central valleys but it takes three days in best weather to go from Tashigang in the east to Ha in the west. Bumthang Valley lying west of Tashigang at an elevation of about 2,956 m is composed of two smaller valleys — Wangdiphodrang on its right and Tashichoeling on its left. The Punakha and Wangdiphodrang forming one valley lies to the north-east of Thimphu at a comparatively lower elevation (1,465 m) through which flows the Pho Chu and Mo Chu or the Sankosh river. The fertile valley of Thimphu lies at a much higher elevation — about 2,600 m — and is of a more undulating nature and much larger in size than Punakha Valley, drained by rivers Wang Chu on which capital Thimphu is situated, and Tashichhodzang flowing southwards through the valley. The widest among the river valleys is the valley of Paro, a perfectly developed plain at an elevation of about 2,135 m, lying west of Thimphu Valley and is the most densely settled area as well as the most fertile of all in the central region of Bhutan, it used to maintain direct trade links with Tibet and India in the past. The valley of Paro is closely followed by another valley in the west, the Valley of river Ha, another fertile though smaller valley than Paro, lying at an elevation of over 2,650 m. It is another fertile valley and almost flat in relief. Among the comparatively smaller valleys are the fertile valleys of Sha (Wangdiphodrang) at an elevation of 1,400 m; the valleys of Mangdi occurring east of Sha at an elevation ot 2,290 m on the banks of river Mangdi Chu; the fertile Tashigang Valley in the far east at an elevation of 1,066 m adjoining the Kurtoe Valley is rich in tropical crops and fruits.

The third physiographic unit, the Southern Foothills Zone also known as the Bhutan Duars, is made up of Gondwanas (Damudas), Buxas or Daling sediments and their respective metamorphics. The Gondwana rocks are composed of quartzite sandstones, including thin layers of coal. It mainly consists of foothills about 1,000 m high but misty, cold hills, some as high as 3,660 m enclose hot, sultry valleys lying at a much lower elevation, the highest of which occur at elevations of 915 to 1,220 m. The remaining area comprises the Duars plains, altogether eleven in number and situated between Tista and Manas rivers they are — Dalimkot, Zumerkot, Chamurch, Lucke, Buxa, Bhulka, Bara, Goomer, Reepo, Chirang and Bagh.

Rivers

The lie of the mountains has given rise to river systems maintaining a southward flow in general. All but two of Bhutan's major rivers begin their long and tortuous courses in the perpetual snows of the enormous glaciers of the northern region of the Great Himalaya while some of them have originated from the lakes in Tibet. All of them having origin in the high mountain or in the high Tibetan Plateau, maintain a torrential course, gushing through extremely narrow gorges in their upper reaches but once the central part of the country is reached, the fury is substantially released where they have formed much wider valleys and flow through them more tamed over the flat floors made up of their own deposits. However, none of them is navigable and even after reaching the plains in India they do not lose the velocity for a while to become navigable. Their hydro-electric potential is enormous and a 336 MW power station, the Chhukha Hydel power generating plant is already functioning on the river Wang Chu.

Among the principal rivers are the Amo Chu, the Wang Chu, the Mo Chu and the Danme Chu most of which ultimately drain through their tributaries into the river Brahmaputra in India.

The Amo Chu or the Torsa river rising in Tangla in Tibet enters Bhutan at Pasha and drains the Chumbi Valley where it maintains a torrential course, making passage through a narrow, deep bed bordered by rocky faces which broadens appreciably on reaching the Duars plains in the extreme south of the country to debouch into the Brahmaputra after crossing Bhutan's territory. Phuntsholing, the principal gateway from India to Bhutan from the south, lies on the bank of the Torsa at the point of its emergence from the mountains.

The Wang Chu, or the Raidak, also rises in Tibet and is joined by the Paro Chu and a few more tributaries like the Do Chu and the Ha Chu etc., the latter rising from the snowy mountains. The Wang Chu is a swift-flowing river known for its sudden fury, maintaining a course of about 370 km in Bhutan before finally entering the plains of West Bengal.

The Mo Chu or the Sankosh river rises in the Great Himalaya above Gasa and joined by its principal tributaries of the Pho Chu and Mo Chu coming respectively from the east and west to get united with it near Punakha Dzong, flows for 322 km in Bhutan. The Mo Chu takes origin in the snowy mountains of Gasa while Pho Chu flows from the mountains in the north-east. The river takes an easterly course and then turns southward and touches the walls of the Wangdiphodrang Dzong flowing therefrom southward under the name of Sankosh and ultimately enters the Duars Plains as a deep, large river to fall into the Brahmaputra, about 48 km above the town Rangamati.

The Dange Chu or Manas river is by far the largest river of Bhutan, commanding nearly the whole of the eastern half of the country as the drainage basin of its combined course. It enters Bhutan from India in the east and receives three important tributaries during its course westward before re-entering India to fall into the Brahmaputra. The principal tributaries of Manas river from east to west are — Dangme Chu, the Kuru Chu and the Mande Chu — of which the last one, the Mande Chu taking origin near Kula Kangri in the extreme north flows southward and becomes known as Tongsa Chu after its course beyond Tongsa Dzong where the level of the river bed is around 1,676 m. The Tongsa Chu is joined further south by the Bumthang Chu before debouching into river Manas on its eastward flow of their combined courses. At this point of confluence with river Tongsa, Manas river flows at an elevation of hardly 122 m while at Tashigang the elevation is about 610 m.

Plate 3 : Museum, Paro

Plate 4 : Tongsa Dzong

Climate

The variations in relief are faithfully reflected in the climate which varies widely roughly from north to south. In general, the three relief or physiographic zones — the Great Himalayan Zone, the Inner Himalayan Mountain and the Valley Zone and the Southern Foothill Zone, define three climatic regions : the Alpine with monsoon, the Temperate with monsoon and the Monsoonic tropical. The alignment of the mountains makes the inflnence of the monsoon felt deep into the north, making it one of the most dominant features of the climate.

In the north, climatic conditions in the Great Himalayan Zone are similar to that of the Tundra with a severe winter and a cold and short summer. The annual rainfall amounts to about 400 mm most of which is concentrated in the summer months, the southwest monsoon accounting for 60 to 90 per cent of the total rainfall.

The Inner Himalayan Mountain and the Valley Zone on the other hand has a contrasting type of climate which in a broad sense is a cool temperate climate characterised by wide variations due to topographic contrasts. For instance, it is a zone of frost experienced particularly at elevations 3,000 m and above. Winters are still severe but summers are cool. The annual rainfall is around 1,000 mm, the western part receiving comparatively higher amount.

The Southern Foothill Zone has a humid sub-tropical climate at an elevation of 900 to 1,200 m which grades into tropical type at lower elevations further south in the Duars. Temperature in the coldest months is less than 10° C and rises to about 24° C in the summer in the Duars. Rainfall varies between 1,200 and 2,000 mm in the foothill areas to 3,000 and above 5,000 mm in the southern fringes of the Duars.

The most characteristic feature of the climate is that variations in temperature as well as rainfall are much wider even between adjoining valleys as determined by topographic features.

However, as aptly described, the climate of Bhutan 'defies any easy description with altitude, rainfall, exposure to sunlight and wind-giving each valley a unique set of conditions'.[3] Ranging from hot and humid tropical and sub-tropical conditions in the south to sub-alpine and alpine types in the snows and perpetual ice of the high Himalaya in the north, the climate presents a set of wide variations in between the contrasting situations of the partly sheltered valley regions and the adjoining rugged terrain.

Nevertheless, altitude seems to be one of the most important factors determining climate so far as temperature is concerned. It may be interesting to note that there are only three districts, namely Samdrupjongkhar (160 m), Sarbhang (210 m) and Samchi (390 m)[4] comprising about 13.98 per cent of the total landmass of Bhutan where the average altitude is below 400 m. In contrast, in nine districts, such as Chirang (1,620 m), Daga (1,520 m), Punakha (1,220 m), Wangdiphodrang (1,260 m), Shemgang (1,916 m), Lhuntshi (1,460 m), Mongar (1,620 m), Pemagatsel (1,200 m), and Tashigang (1,040 m), the average altitude is over 1,000 m but less than 2,000 m. Together, these nine districts comprise about 57.31 per cent of the total area. The remaining six districts of Thimphu (2,320 m), Chhukha (2,220 m), Ha (2,712 m), Paro (2,280 m), Bumthang (2,690 m) and Tongsa (2,180 m) together representing almost 28.71 per cent of the land area lie at an altitude exceeding 2,000 m. Thus, barring Tashigang, fourteen out of a total of eighteen districts containing nearly 76 per cent of the total area of the country lie at an elevation over 1,200 m.

This has a direct impact on temperature regime in the country as reflected in the very low temperature recorded even in the enclosed valley regions of Thimphu, Paro and the like. At Thimphu, for instance, the minimum temperature drops to freezing point as early as in November, going below it in December and remains near freezing point in all the successive three months with 0.4°C, 1.0° C and 1.1°C for January, February and March respectively. Bondey in Paro valley provides a similar instance where the temperature remains below freezing point from November (-0.2°C) to December (- 0.4°C) and possibly in January too for which data is not available, continuing to be near freezing in February (0.2°C) and March (0.8°C). At Ura, in Bumthang district, the mercury dips below freezing point and the nights are bitterly cold with a minimum temperature remaining at that level for the whole of November and continuing perhaps till April though data for January to March is not available. Perhaps, Lingshi lying in the north of Thimphu district holds the lowest record of minimum temperature in the country with mercury dropping below freezing point all through January to April and probably in the earlier months of November and December as well for which data is not available. February records the lowest minimum with -8.7°C. The day temperature during the winter months as well remains quite low going below 10°C. At Lingshi, it is 1.5°C in February. At Namjeyling in Ha district, data available for February and March shows 7.6°C and 7.9°C respectively as maximum temperature. At Nobding and Phubjikha in the district of Wangdiphodrang, the day temperature in January and February remains at 9.6°C and 9.3°C in the former and 8.8°C and 7.4°C in the latter.[5]

On the whole, November to February is the coldest period in Bhutan but incomplete data for many stations and the fact that "temperature data for some stations may not be very accurate" make it rather difficult to get a full picture. However, as the available data shows, the summer is at its height during the months of June to August in general, and June is the hottest month recording the highest temperature for 32 out of 68 stations, closely followed by August with 26 stations experiencing the second highest maximum temperature. May, July and September are quite hot for many stations before or after which day temperature drops visibly. The highest maximum day temperature of 35.3° C in June is recorded for Yebulapcha in Shemgang district followed by the next highest of 33.4° C in June at Bakuli in Samdrupjongkhar district. In general, the higher day temperatures are experienced in the southern districts though it may be as high as 31.5° C at Simtokha in Thimphu district. Besides, as it is recorded in some stations, like Shemgang, Thrimshing, Langthel, Panbang, Tongsa, Ura, Kanglung, Yallang, Yadi, Dungamain, Daifam, etc., the day temperatures during the summer months remain almost uniform for which altitude, exposure and the effect of rainfall seem to be responsible.

Table 2.1 : Day Temperature in Bhutan (in Degrees Celsius)

District or Dzongkhang	Station	May	June	July	August	September
Thimphu	Thimphu Agricultural Office		27.0	26.9	27.2	
Chhukha	Chhukha		n.a.	24.3	24.5	24.3
	Gedu		19.2	18.6	19.4	20.3
Paro	Bitekha		20.7	20.6	20.5	
	Bondey			24.4	24.6	24.7
Samchi	Samchi		28.6	27.2	28.3	n.a.
Chirang	Damphu		21.3	21.3		
Daga	Dagadzong	22.5	24.2	22.6	22.5	
Punakha	Gasakhatey	20.2	18.2	20.3	20.3	
Wangdiphodrang	Gaselo		25.0	25.6	25.4	25.1
Sarbhang	Sarbhang	30.5	29.0	n.a.	30.4	n.a.
Shemgang	Shemgang		22.5	22.4	22.8	22.8
Tongsa	Langthel	28.2	28.7	29.1	29.9	29.3
Bumthang	Bumthang		19.1	18.1	18.1	
Tashigang	Thrimsing		27.1	27.2	27.7	27.9
	Yallang		20.7	20.2	20.4	n.a.
Mongar	Yadi	24.2	25.2	24.9	25.7	24.7
Lhuntshi	Dungkhar		25.2	23.6	24.1	
Pemagatsel	Dungamain		24.7	24.1	24.5	24.5
Samdrup-jongkhar	Daifam	30.5	30.5	31.3	31.6	30.5

Source : Statistical Year Book, 1990.

The wide variability of annual rainfall derived from the average for seven years is better noticeable from its regionwise distribution. For instance, the Eastern Region of Bhutan receives very little rainfall, with Kurijampa receiving the lowest amount of about 450 mm in the country. The amount increases to about 752 and 800 mm at Tashigang and Tangmachu respectively, taking the fifth and sixth position among the lowest rainfall areas in the country[6] (Chhetri, 1992). It comes mostly during the monsoon period, the highest being between June and September while the winter months of November to February often go dry. The annual rainfall is also low in the Western Region, Thimphu (592 mm),

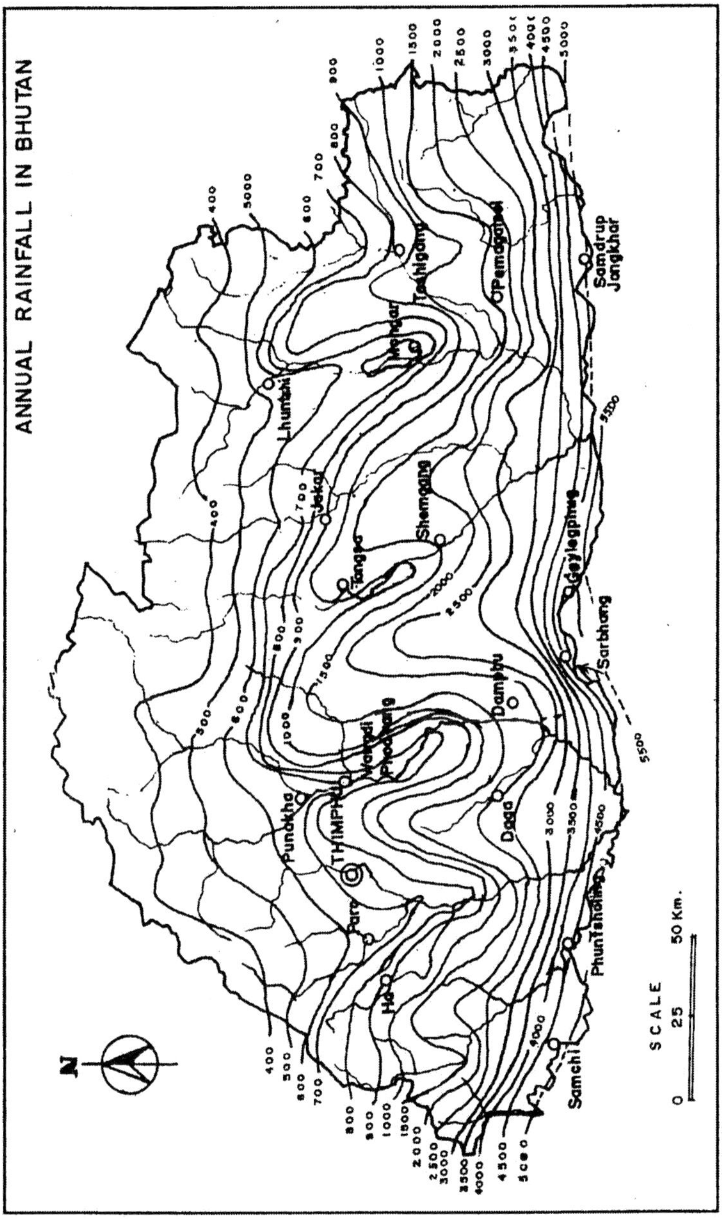

Fig. 2.1 : Annual Rainfall in Bhutan

Wangdiphodrang (627 mm), Paro (685 mm) and Ha (870 mm) respectively take second, third, fourth and seventh place in terms of the lowest record of average annual rainfall, the highest of which in this region occurs at Taba (940.7 mm). The winter months are generally deprived of rain which is also quite negligible during the pre-monsoon months followed by the months of June to September receiving the major part of the rainfall. In contrast, there is a sharp increase in rainfall in the central and the interior belt of the Southern Region, ranging between 1,001 and 2,000 mm taking place, as recorded, at places like Tongsa, Punakha, Chhukha, Shemgang, Dagapela etc. with almost a dry winter. The amount increases further in the Southern Region, having the highest record of an annual average of 6,018 mm at Bhur in Sarbhang district closely followed by Kalikhola (4,826 mm) in the same district and Gedu (4,774 mm) in Chhukha district. A high amount of rainfall is also reported from Arong (4,672 mm) in Samdrupjongkhar district, Surey (4,513 mm), Sarbhang district, Pakbang (4,456 mm) of Shemgang district and in Dorokha (4,214 mm) of Samchi district (Fig. 2.1). The intensity of the concentration of rainfall is remarkable for certain stations. Ordinarily, the major part of the precipitation occurs in June and July though there are exceptions.[7]

Nevertheless, the distributional pattern of rainfall in all the four regions has a marked resemblance such as July is the peak rainy month followed by June and August, having the second peak period with further decrease in September but the amount of rainfall till then is higher than in May, petering out afterwards and bringing in nearly a dry period beginning from November onwards lasting till February.

REFERENCES

1. *Geology and Mineral Resources of Bhutan,* November 1977, Geological Survey of India, Bhutan Unit.
2. *Ibid.*
3. *A World Bank Country Study: Bhutan,* Development in a Himalayan Kingdom, 1984, The World Bank, Washington D.C., p. 2.
4. *Statistical Year Book of Bhutan 1990,* Planning Commission, Royal Govt. of Bhutan, Thimphu 1991.
5. *Ibid,* Table 5.4, p. 30.
6. Chhetri, D.B., December 1992, Forest Fire Season in Bhutan in *Tsenden : A General Publication on Forestry in Bhutan,* The Department of Forestry, Royal Govt. of Bhutan Thimphu, Vol. 4, No. 1.
7. *Statistical Year Book of Bhutan 1990,* Table 5.5.

3

MINERALS AND POWER

— *Bimalendu Bhattacharya*

Minerals

THE mineral resources of a country have an intimate relationship with its geology and in Bhutan a number of minerals of economic value are associated with it. Small scale mining activities may be traced to a period of over a thousand years in Bhutan. The presence of slags in many parts of the country indicates the mining of lead-zinc and iron ores for the manufacture of 'artifacts, weapons, and most probably a series of iron chain suspension bridges ... constructed in the 14th century by saint Thangthong Gyalpo (1385-1464)'.[1] However, mineral exploration in real sense began only in the early 1960s and less than 30 per cent of the total area of the country so far explored and geologically mapped, have revealed the presence of a number of important minerals.

Beryl is one of them occurring in parts of Wangdiphodrang- Rukubji, Tongsa-Khosala, Ura-Gajam Chu and Chirang- Domphu areas.[2] *Calc-Tufa* occurs in the Kalapani area of Samchi district with an estimated cement-grade reserve of 1.04 m tonnes. *Coal* forms an important mineral occurring in a 75 km long belt of Gondwana Formation between Kherkheria in the east and Diu Ri in the west in the southern parts of Samdrupjongkhar, Tashigang and Sarbhang districts of eastern Bhutan. Detailed exploration has been carried out in Gerua and Dimala Khola blocks and a total reserve of 89,000 m tonnes of coal has been estimated at 50 m depth from the surface at Gerna block. A small amount of coal is mined in Bhangtar in Samdrupjongkhar which is exported to the nearby tea gardens in India and also to Bangladesh. The total reserves of all grades of coal upto a depth of 20 to 30 m are estimated at 1,032 m tonnes. There are 11 important *dolomite* deposits in the foothills close to Indo-Bhutan boundary located at Samchi, Sarbhang and Samdrupjongkhar districts having a total reserve of 10,525 m tonnes upto a depth of 30 m. The quality of dolomite is very good which is suitable for use in steel, glass and manganese metal manufacturing processes. It is exported to India and Bangladesh. Two important occurrences of *graphite* have been found on one slope of the Khep-chishi hill and Chilai La in Paro and Ha districts with proved reserve of 79.15 m. tonnes of which 5 m. tonnes appear to be of better quality. A total reserve of about 70 m. tonnes of *gypsum* has been traced, the major portion (56.45 m. tonnes) of which occurs at Khothakpa village in Samdrupjongkhar district. The other two deposits are at Uri Chu and Khar near Khothakpa. Gypsum is mined in Pemagatsel for use in the cement plants. Some is exported to India. *Lead-Zinc* occur in small deposits in Thimphu, Samchi, Samdrupjongkhar and Ha districts of which that in Thimphu is the most important one. The total reserve is estimated at 0.585 m. tonnes. *Mica* occurs in book form in five pegmatite horizons in the granitic gneiss in Chirang district. A number of important deposits of *Limestone* have been traced in Samchi, Paro, Samdrupjongkhar, Sarbhang, Tashigang and Wangdiphodrang districts. A deposit of about 1.94 m. tonnes of cement-grade limestone has been found at Rongri area in Sarbhang district. There is also a deposit of 4.012 m. tonnes of cement-grade limestone in the Pugli extension area and another reserve of 0.9 m. tonnes has been located at Titi area. A reserve of 2.8 m. tonnes has been traced in the Kangrazhi and Kurung Ri area of Nanglam. Limestone is mined for use in the cement and calcium carbide manufacturing plants in Bhutan. A total reserve of cement-grade limestone has been estimated at 118.54 m. tonnes. A number of important *marble* deposits have been traced at Khanku and Paro in Paro district, Ha in Ha district, Genekha in Thimphu and Tala in Samchi districts together accounting for a total reserve of 516.66 m. tonnes.[3] *Talc* has been located in small deposits in Samchi and Sarbhang districts with a total reserve of 28,500 tonnes and *Steatite* occurs in Samchi district with a proved reserve of 1,392 tonnes. *Slate* occurs in Wangdiphodrang district, having a reserve of 16 m. tonnes upto a depth of 20 m. Besides, deposits of *iron ore, tungsten, phosphate, asbestos* and *gold* have been traced. Investigations are on in Gongkhola, Tongsa district for *copper* deposits. Drilling in Burkhola in Sarbhang district for strategic minerals has revealed a reserve of 3.42 m. tonnes of *scheelite* ore.

The findings are the results of mapping of over 25,000 sq. km. area of the country, leaving information waiting for nearly another half representing the most complicated, largely inaccessible areas of the northern and the north-eastern parts. The minerals described above are not large enough in many cases for commercial operations. Besides, the terrain is difficult in many areas and the overall absence of infrastructural facilities have kept industrial prospect very much limited, making it rather a compulsion for the country to seek compensation for this lack in other fields, such as agriculture for bolstering up its economy.

Power

It is aptly said that power is progress since power in the form of electricity is the prime mover in all activities concerned with economic (and social too) development of a country involving industry, agriculture, transport, communication etc. as well as both domestic and public life at all stages. The vast forest resources of Bhutan has traditionally been the major source of firewood, constituting the major energy source for cooking and heating where the transition from bio-fuels (wood, dung, corn residues etc.) has rather been a slow process. The rate of consumption of fuelwood is determined by the eating and the social habits of people, and accordingly rural Bhutan has broadly been divided into two zones — one comprising six districts (Dzongkhags), namely Tashigang, Mongar, Pemagatsel, and Shemgang as the major 'ara'(the local wine) producing areas for private consumption, and Daga and Tongsa as the minor producing areas, in all of them maize and rice being the staple food while in the remaining districts the staple food is rice and wheat where people generally favour tea as the primary drink (Raina, 1991).[4] The consumption of firewood in the 'ara' producing areas comes to about 6.7 kg per capita per day whereas in the 'non-ara' areas it is about 3 kg. In other words, the per capita consumption of energy amounts to 0.76 tonnes of oil equivalent (TOE), but as most of it is firewood, it amounts to 0.13 (TOE) due to the low efficiency of firewood burning technology. Firewood alone accounted for nearly 84 per cent of the useful energy consumed at the beginning of the Fifth Five Year Plan (1981-87). However, the dependence on forests as the source of firewood started diminishing slowly in the later years with the awakening of great interest in checking the depletion of forests. Thus, in 1989, firewood accounted for 77 per cent of the total energy consumption or an estimated 1,170,000 cubic metres required for all the different users of firewood like households, commercial establishments such as agro-industries, institutes such as schools, monastic bodies, army, police etc. The households or the domestic sector used 95 per cent of the total consumption of firewood in the country, the Government and commercial establishments 3 per cent, the agricultural sector 0.9 per cent and the industrial sector used only 0.7 per cent. For a typical household the annual consumption came to about 14 cubic metres of which 7 to 9 cubic metres was consumed in cooking and for house heating.[5]

The annual fuelwood consumption in the rural areas of Bhutan for cooking and the maufacture of 'ara' was 2333.3 million kg and is expected to amount to about 2843.4 million kg in around 2000.[6] Eucalyptus, leucaena, poplar and willow have been recommended to be grown for supplying the domestic fuel, fodder and sometimes building wood. Among them, eucalyptus, a native tree of Australia called 'gum tree' or 'string bark', has become quite popular in the dry areas of Wangdiphodrang and Tashigang.[7] Besides, forage grasses and legumes, as recommended, could be a large source of biomass which could be used in gasifiers for producing gas for cooking and heating.

Though there is very little coal and no petroleum deposits, Bhutan has a vast potential in hydel power tied up in its scores of rivers coursing through mountain terrain, the exact amount of which has not yet been fully ascertained because of incomplete survey. Theoretically, it has been estimated at 20,000 MW although the exploitable capacity is estimated to veer round 6,000 MW.[8]

The first micro hydro-electric power plant became operative in Thimphu in 1967 with a capacity of 360 kW and from this modest start, power generation went up to a total of 355 MW at the beginning of the Seventh Five Year Plan (1992-97). The steady increase in power generation from 1967 onwards was a result of the growing demand of the domestic, industrial and the agricultural sector which is clearly reflected in the change in the consumption pattern through the successive plan periods. For instance, while 80 per cent of the electricity generated by the power plants

Plate 5 : Prototype of improved rural house with traditional architecture and modern amenities.

Plate 6 : Tongsa road-side hotels-cum-residence and shops.

was used by the domestic sector alone at the end of the Third Plan, by 1981-82, that is, at the beginning of the Fifth Plan it came down to 45 per cent for the domestic sector when 45 per cent was consumed by the industry and commerce, and 10 per cent went into use for the agricultural sector. At the same time, the growing consumption in all the sectors made it necessary to import power from the neighbouring country, India, to the extent of almost 60 per cent of its total requirement.[9]

In order to narrow down the widening gap between demand and supply of power, particularly keeping in view the proposals for the industrial programme during the Fifth Plan period, the Chhukha Hydel Project was conceived at the end of the Fourth Five Year Plan, the commissioning of which was likely to solve the problem besides leaving a comfortable surplus as a direct source of revenue from its sale to India providing a ready market for years to come.

The most important achievement in regard to the production of power on a large scale basis was the commissioning of the Chhukha Hydel Corporation during the Sixth Plan period with an installed capacity of 336 MW and the revenue earning from the sale of power at once shoot up to Nu. 375 miilion a year in 1991. No account of power development in Bhutan is complete without a reference to the Chhukha Hydel Power Project, the largest of its kind in the country, making first major initiative towards exploiting Bhutan's enormous power potential, harnessing the waters of the Wang Chu river in western Bhutan over a drop between Chimakothi and Chhukha. Located at a site between 83 and 97 kilometres north of Phuntsholing on the road to Thimphu, the project involved the construction of a 40 m high gravity dam to divert the river waters after desilting into a headrace tunnel and taken to an underground powerhouse with four 84 MW turbines. The construction started in 1973 and was completed in 1988.

The number of high voltage transmission lines and sub-stations established following the completion of the Chhukha Project covered the western part of the country with a well-laid grid extended further by the construction of 119 km of transmission line by the end of 1990 establishing connection between the following places: Phuntsholing-Pasakha; Phuntsholing-Penden; Chuzam-Paro; Chuzum-Ha; Simtokha-Wangdiphodrang. During this period, a major work was done in Thimphu where most of the sub-transmission and distribution network were placed underground and overhead lines were made more stable. The rural electrification programme made significant progress during the Sixth Plan period, increasing the number of electrified villages from 127 in 1986-87 to 171 in 1988-89 and the number of towns electrified increased from 19 to 20 while the number of consumers rose from 11,361 tc 14,092[10], though till then over 90 per cent of the population went without electricity for personal consumption.

The power position strengthened further during this time with the completion of 12 mini and macro hydel plants to be followed by the completion of three more micro hydel plants within the Seventh Five Year Plan period. The micro hydel plants completed were as follows :

1. Khaling Mini Hydel Plant with a capacity of 3 x 0.2 MW.
2. Gyetsa Mini Hydel Plant with a capacity of 3x 0.5 MW.
3. Thinley Gang Micro Hydel Plant with a capacity of 30.0 kW.
4. Tamshing Micro Hydel Plant with a capacity of 30.0 kW.
5. Surey Micro Hydel Plant with a capacity of 70.0 kW.
6. Kekhar Micro Hydel Plant with a capacity of 20.0 kW.
7. Bubja Micro Hydel Plant with a capacity of 30.0 kW.
8. Tongsa Micro Hydel Plant with a capacity of 50.0 kW.
9. Rukubji Micro Hydel Plant with a capacity of 40.0 kW.
10. Tansibji Micro Hydel Plant with a capacity of 30.0 kW.
11. Ura Micro Hydel Plant with a capacity of 50.0 kW.
12. Yadi Micro Hydel Plant with a capacity of 30.0 kW.

The plants to be completed were :

1. Dagana Micro Hydel Plant with a capacity of 200 kW.
2. Damphu Micro Hydel Plant with a capacity of 200 kW.

Apart from hydroelectric power other sources of energy play a very insignificant role in power generation in Bhutan. The petroleum products like diesel oil, kerosene, furnace oil, LPG etc. imported from India accounted for 8 per cent of the energy consumed in the country in 1989. Coal, with an annual production of 20,000 metric tonnes (1990), accounted for 0.8 per cent of the total energy consumption used particularly by such industries as food processing units, lime kilns, candle and brick factories. It has to be noted here that coal is both imported as well as exported. For instance, in 1988, about 18,000 metric tonnes of coal was imported from

India while in the same year 19,000 metric tonnes was exported to Bangladesh. Further, 5,000 metric tonnes of coke was imported from India in 1988. The share of biogas is less than that of coal, produced by a total of 54 plants in Sarbhang district, each with a capacity of 2 to 3 cubic metres per day used only for lighting purposes. There are two more plants in Punakha district, one at Ritsha having a production capacity of 4 cubic metres per day and another in Lakha with a capacity of 9 cubic metres per day.

The Department of Power has provided 271 solar panels to some of the monasteries, institutions, Basic Health Units in remote areas used for charging the batteries for lighting purpose though some of them may not be working. Wind energy is used for charging the batteries at two places : in the Swiss Farm in Jakar, Bumthang district, and in the Centre for Agriculture Research Development (CARD) in Wangdiphodrang for irrigation.

As it appears, power generation is absolutely dominated by two sources : the hydel power and the diesel power of which the former enjoys an undisputed supremacy considered even in terms of the total value of the power generated in Bhutan (Table 3.1). A number of interesting features emerge from this Table, such as : (i) the role of the Department of Power in power generation has declined steadily through the years from a share of 2.02 % in 1986 to 0.16% in 1989 of the total value of power produced in the country; (ii) a similar trend is observed in the case of diesel, sliding

Table 3.1 : Power Generation by the Department of Power and the Chhukha Hydel Project in Bhutan.

(Diesel shown as % of the total)

Department	1986	1987	1988	1989
Department of Power				
(a) Hydel	5.452	5.38	2.711	2.527
(b) Diesel	1.546 (0.446)	0.313 (0.021)	0.239 (0.015)	0.022 (0.001)
Sub-total	6.998 (2.02)	5.693 (0.39)	2.95 (0.19)	2.549 (0.16)
Chhukha Hydel Corporation	339.386 (98.00)	1470.466 (99.61)	1541.057 (99.81)	1554.301 (99.84)
	346.384	1476.159	1544.007	1556.85

Source : Seventh Five Year Plan (1992-1997), Planning Commission, Royal Government of Bhutan.

down gradually from a share of 0.446 per cent of the total value of power in 1986 to a mere trifle of 0.001 per cent in 1989; (iii) the value of power produced by Chhukha Hydel Corporation meanwhile has maintained an upward trend, increasing from 97.98 per cent in 1986 to 99.83 per cent of the total value of power generated in Bhutan in 1989, and during this period Chhukha has multiplied its own contribution nearly five-fold.

The Royal Government of Bhutan considers the power sector as one of the principal sources for earning revenue through the internal sale of power as well as export of power to India which alone provided Nu. 384 million in 1991. This is likely to increase further in near future with the completion of Kurichu Hydel Project (45 MW), Tangsibji Hydel Project (45 MW) and Basachu Hydel Project (41 MW).

Determined on the basis of economic and technical viability, the main objectives of the Seventh Five Year Plan are to :

(a) increase Government revenues through the generation of power for sale to India and to industries in the country, and
(b) fulfil the demand for electrical energy at minimum cost and distribute it through an efficient as well as reliable system in a uniform manner as far as practicable to the people of Bhutan almost 90 per cent of whom did not have access to it;
(c) ensure a balanced regional development through the provision of energy for industrial growth, and
(d) ensure an environmentally sustainable use of water power through watershed management.[11]

Among the major strategies to be adopted for achieving these targets it is mentioned in the Main Plan Document of the Seventh Five Year Plan that the implementation of Tangsibi Hydel Project in central Bhutan, Kurichu Hydel Project, Rangjung and Yonglachu Hydel Projects in Eastern Bhutan are expected to contribute largely to attaining regional balanced growth. Besides, investigations for two large projects on Chhukha, namely Chhukha II and Chhukha III were to be completed for a generating capacity of 1,000 MW and 600 MW respectively. The Power Master Plan becoming operative in 1991, was to provide a list of potential hydel power project sites to be investigated for final decisions.

REFERENCES

1. Seventh Five Year Plan 1992-1997, Vol. I, Main Plan Document, Planning Commi- ssion, Royal Govt. of Bhutan, December 1991.
2. *Geology and Mineral Resources of Bhutan,* November 1977.
3. *Minerals of Bhutan,* Samchi 1979, Geological Survey of India, Bhutan Unit.
4. Raina, B.N., 1991, Energy, Environment and Bhutan, in *Bhutan and Its Natural Resources,* Sherabtse College, Kanglung, Vikas Publishing House Pvt. Ltd., Delhi.
5. Seventh Five Year Plan 1992-1997, Vol. I, Main Plan Document.
6. *Op. cit.*
7. *Druk Forestry News,* June 1993, Ministry of Agriculture, Royal Govt. of Bhutan, Vol.1, No. 2.
8. Seventh Five Year Plan 1992-1997, Vol. 1, Main Plan Document.
9. Bhutan, Country Economic Memorandum, May 1983, Vol. I, Main Report. Prepared for the Asia and Pacific Round Table Meeting, Geneva, Planning Commission, Royal Govt. of Bhutan, p.24.
10. Seventh Five Year Plan 1992-1997, Vol. I, Main Plan Document.
11. *Ibid.*

4

AGRICULTURE

— *Bimalendu Bhattacharya*

AGRICULTURE is the dominant sector of economy contributing directly about half of GDP, accounting for over one-fourth of export earnings, and providing an employment base to as much as 95 per cent of the total population.[1] In the absence of adequate climatic data, altitude has been taken as the determining factor to divide Bhutan into agro-ecological zones[2], which means areas having more or less similar ecological conditions giving rise to similar production systems related to crops, livestock and forests. Such ecological conditions include topography, soil, rainfall, temperature and natural vegetation (Fig. 4.1).

From agricultural point of view, the land has been classified in terms of its capability based on such considerations as slope, soil depth, soil texture, stoniness, drainage, previous erosion and altitude. In other words, it is the qualitative determination of the limitations of an area of land for fruitful use, such as for agriculture, pasture or recreation. On this basis, 10 classes of land have been defined[3] of which land falling under class I to IV is capable of supporting agriculture or horticulture and among them Class I is considered to be the best while class V is marginally suitable for such purposes. On the other hand, land classed as VI and VII is not suitable for agricultural use due to severe physical limitations like steep to very steep slope, very shallow to stony soils, a considerable area of rock

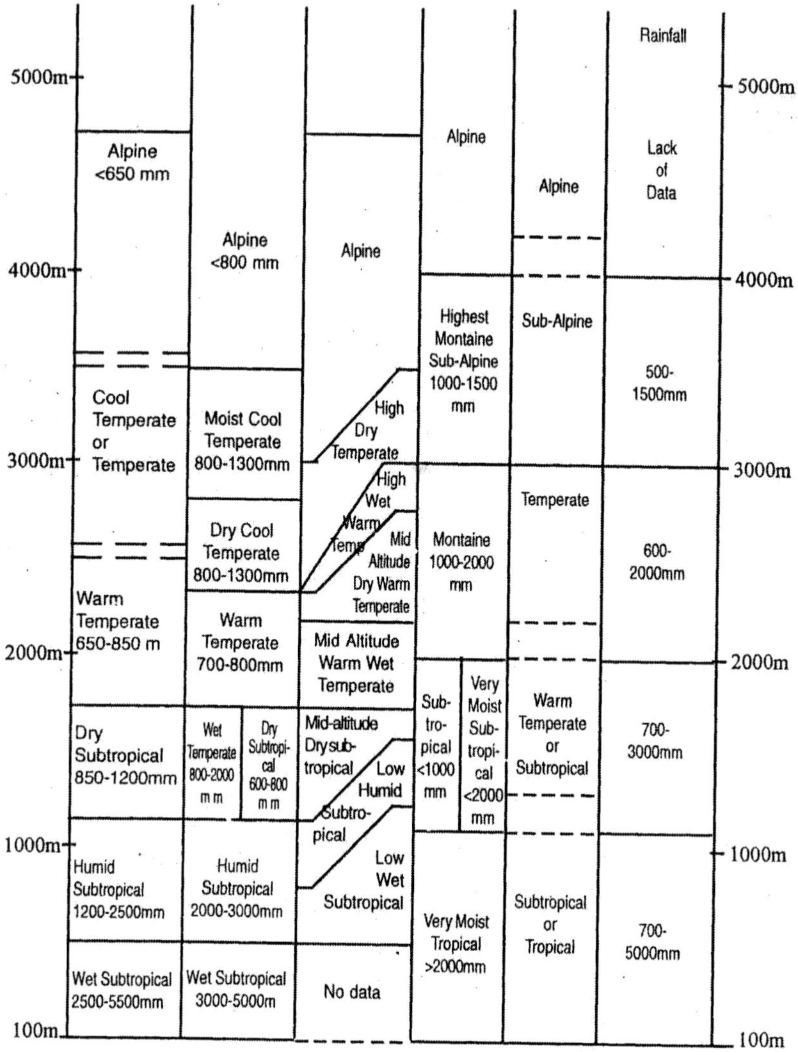

AGRO-ECOLOGICAL ZONES OF BHUTAN

Source : LUP *Newsletter*, Vol. 1., No. 3, October, 1993.

Fig. 4.1 : Agro-ecological Zones of Bhutan.

outcrops, or a high risk of erosion. In contrast, classes VIII, IX and X comprise land which because of altitude or very unfavourable physical conditions, cannot be put to any type of agricultural or horticultural use except that some of it may be found useful for yak grazing. The classification made under the initiative of LUPP (Land Use Planning Project) has further been renamed as 'Land Potential Classification' so that it can be used both for land already used and land not used so far and according to LUPP, this is the only way to understand the actual character or potential of the land since Land Capability Classification is normally undertaken for land not under agricultural use at present, that is when new land is to be taken into consideration.

A mountainous country like Bhutan has its limitations in arable land, no doubt, and the situation is further complicated by abuse or uneconomic use of available land. An idea about the paucity of agricultural land and the nature of present land use may be obtained from the following data (Table 4.1).

Table 4.1 : Land Use in Bhutan *(Area in sq km)*

	Type and Sub-Type	Area	as % of the Total
1.	Forest	29,045	72.5
2.	Pasture		
	(a) Natural Pasture	1,553	3.9
	(b) Improved Pasture	11	0.0
3.	Agriculture	3,088	7.7
4.	Horticulture	58	0.1
5.	Settlements	31	0.1
6.	Others	6,289	15.7
	(a) Snow/Glaciers	2,989	7.5
	(b) Rock Outcrops	2,008	5.0
	(c) Water Spreads	304	0.8
	(d) Marshy Areas	35	0.1
	(e) Landslips/Erosion	954	2·4

Source : LUP Newsletter, Vol. 3, No. 1, January 1995.

According to available data, forests including natural and improved pastures come to over 76.4 per cent of the total land area of the country where agriculture along with horticulture occupies only about 7.8 per cent, and barely 0.1 per cent of the land comes under the settlements. On the other hand, almost 15.7 per cent of the total land remaining under perpetual snow or glaciers or with rock outcrops and comprising landslip or erosion prone areas, are not suitable for any economic purpose.[4]

There has, however, been changes in land-use initiated during the period of the Five Year Plans, beginning from 1961, and agriculture received greater attention as nearly 95 per cent of the population of Bhutan are dependent on agriculture and the contribution of this sector including animal husbandry alone to GDP was 55.7 per cent in 1980-81. Between 1958 and 1989, the most detectable change has occurred in the agricultural area, having registered a significant increase, nearly doubling in amount with a major shift taking place in the broadleaved forest areas which is getting slowly converted into agricultural and grazing lands in answer to meet the requirements of the growing population. At the same time, land under degraded forests has also increased, though at a much slower rate indicating unwanted destruction of the forests. In contrast, land under the coniferous forests has remained almost stable (Table 4.2). The decrease in forest cover has been estimated at 16.7 per cent of the 1958-coverage and 34.3 per cent over hundreds of years.

Table 4.2 : Changes in Land-Use in Bhutan 1958 – 89

Class ('000 ha)	1958	1989
Agriculture	299	559
Broadleaved Forests	1,485	1,056
Coniferous Forests	1,011	1,006
Degraded Forests and Other	1,245	1,426
Total	4,040	4,042

Source : Seventh Five Year Plan, 1992-97, Royal Government of Bhutan.

Land Under Cultivation and Farming Systems

The land under cultivation comes under different categories determined by physical conditions with a varied degree of limitations giving rise to a wide variety of crops resulting from different types of farming systems. Cultivation in Bhutan is practised in five different ways (Table 4.3) : they are — (a) valley cultivation; (b) terraced cultivation; (c) unterraced cultivation; (d) shifting cultivation; and (e) orchards. Among them the largest land occupance is held by unterraced cultivation, comprising over

46.9 per cent of the total cultivated land while second in rank comes the terraced cultivation. Shifting cultivation, known as Tseri in Bhutan still holds, a significant place, taking up more than 11.57 per cent of the total cultivated land.

Table 4.3 : Land Under Cultivation in Bhutan 1991

Type	Area (in hectares)
Valley Cultivation	46,000
Terraced Cultivation	204,000
Unterraced Cultivation	304,000
Shifting Cultivation	75,000
Orchards	19,000
Total	648,000

Source : Seventh Five Year Plan 1992-97, Royal Government of Bhutan.

A distinctive picture of farming systems has developed in Bhutan consistent with its physical background and associated agricultural and related activities. Thus, each farming system is, in many ways, different from the others, reflecting human response to environmental settings. Each farming system has developed on the basis of different activities carried out by farming households "within their physical and socio-economic environment, to produce food and to satisfy other needs such as fuel, clothes, medicines, cash income and so on. There are different farm systems, each characterised by a particular set of management practice and techniques adapted to the soil, climate, topographical and hydrological conditions". While there is too much emphasis on the main crop in a farming system, the most characteristic element of it is the mixed character, integrating agriculture , livestock and forestry. A glimpse into the life style of the farm households may be obtained from the study of any farming system. In fact, they are manually operated mixed agriculture/livestock systems working mostly on a subsistence level, although cash crops are gradually becoming more important products of the farming system. As a rule, most of the farmsteads possess some heads of animals and maintain a kitchen garden besides some fruit trees or a small orchard or plantation. The domestic animals are grazed in the pastures or adjoining forest land. The farm yard manure in the temperate zone and tethering of animals in the tropical zone contribute to restoring soil quality.

A broad classification of the farming system based primarily on the nature of land identifies five types :

Wetland Dominated Farming System

It is dominant mostly in the west and south of the country where paddy is the main crop followed by wheat, potatoes and vegetables. Some amount of dry farming and/or tseri (shifting cultivation) is also practised here and small plots of kitchen gardens maintained. In eastern Bhutan, the farmers keep and protect little oak forests in between cultivated areas for manure. These are called 'sokshings'. Domestic animals like cattle, pigs and poultry for draught power, meat and eggs make a common part of the farm households.

Dryland Dominated Farming System

It is not specifically confined to any part but occurs throughout the country wherever rainfall is short of the requirement of the major cereal crops. Potatoes and wheat are the main crops in the temperate zone followed by mustard, buckwheat, turnips and vegetables. In the sub-tropical zone, maize and millet become the most important crops. Everywhere the farmers keep animals like cattle, pigs and poultry as major sources of food as well as economic support.

Mixed Wetland and Dryland Farming System

For the farmers in Bhutan both wetland and dryland farming carry similar weight and crops as well as farming practices are not much different from that of the Wetland Farming System . However, the constraints seem to be quite different, particularly "in areas where wetland and dryland are far apart and sometimes even separate houses are maintained at both sites."[5]

Tseri Dominated Farming System

It is prevalent in the central, eastern and southern hill areas. Here maize is the main crop and among other crops millet, buckwheat and rice take an important place. The farmers also rear some animals. The majority of the farmers own or share small areas of wetland and dryland. The farming practices under Tseri or shifting cultivation are quite different from that of wet or dryland practices.

Pasture/ Livestock Based Farming System

Occurring mostly in the north, agricultural practice is very insignificant in these areas or virtually absent. Yak and sheep are kept in the Alpine zone while cattle at lower altitudes are kept by farmers practising agriculture and only a limited number of them maintain a large herd. The herders are in the habit of practising transhumance, taking herds to lower elevations in the winter months in search of pasture and fodder. As a rule, animal products are bartered for other food and necessary items.

These major farming systems may further be sub-divided based on physical conditions, bringing out greater disparities in crop pattern and life styles.

Cultivation Practices and Opportunities

The abject paucity of land under agricultural uses is further emphasised by its districtwise distribution (Table 4.4). The country is divided into 18 districts (Dzongkhag) and Samchi alone contains 16.7 per cent of the total cultivated land while Samchi along with Chirang (10.8 %), Geylegphug (12.2 %), Tashigang (13.7 %) and Samdrupjongkhar (7.8 %) in the south maintain 61.1 per cent of the total land put to different types of uses, like wetland, dryland, orchards, tseri and kitchen garden (excluding land under the head 'others' in Table 4.4). In spite of having wide valley floors like Paro, Thimphu, Punakha and Wangdiphodrang, land under agriculture does not exceed 2.9 per cent in Paro and much less in other three because of rough terrain.

Table 4.4 : Land Under Agricultural Use by District or Dzongkhag in Bhutan 1988-89

(% shown within brackets) *('000 hectares)*

District or Dzongkhag	Wet Land	Dry Land	Tseri	Kitchen Garden	Orchard/ Plantation	Others**	Total
1	2	3	4	5	6	7	8
1. Thimphu	800 (3.0)	500 (0.9)	100 (0.3)	—	—	6,900 (5.2)	8,300 (3.2)

(Contd.)

Table 4.4 : *(Contd.)*

1	2	3	4	5	6	7	8
Zone I							
2. Chhukha	1,100	3,100	4,400	—	700	8,200	17,500
	(4.1)	(5.6)	(13.4)		(7.1)	(6.2)	(6.8)
3. Ha	100	800	300	—	300	8,200	9,700
	(0.4)	(1.5)	(0.9)		(3.0)	(6.2)	(3.8)
4. Paro	1,500	1,600	200	100	200	3,400	7,000
	(5.6)	(2.9)	(0.6)	(11.1)	(2.0)	(2.6)	(2.7)
5. Samchi	4,800	7,900	4,300	200	3,700	2,300	23,200
	(17.9)	(14.4)	(13.1)	(22.2)	(37.5)	(1.7)	(9.0)
Zone II							
6. Chirang	3,200	7,200	600	300	2,200	400	13,900
	(11.9)	(13.1)	(1.8)	(33.4)	(22.2)	(0.3)	(5.4)
7. Daga	800	2,600	800	—	200	2,700	7,100
	(3.0)	(4.7)	(2.4)		(2.0)	(2.0)	(2.8)
8. Punakha	1,600	400	—	—	—	8,300	10,300
	(6.0)	(0.7)				(6.3)	(4.0)
9. Wangdi-phodrang	1,300	900	300	100	—	na	2,600
	(4.8)	(1.6)	(0.9)	(11.1)			(1.0)
Zone III							
10. Bumthang	—	1,100	1,900	—	—	14,100	17,100
		(2.0)	(5.8)			(10.8)	(6.6)
11. Geylegphug	4,800	7,500	900	200	1,800	300	15,500
	(17.9)	(13.7)	(2.7)	(22.2)	(18.2)	(0.2)	(6.0)
12. Shemgang	700	1,600	3,300	—	300	5,600	11,500
	(2.6)	(2.9)	(10.2)		(3.0)	(4.2)	(4.5)
13. Tongsa	700	800	800	—	200	4,000	6,500
	(2.6)	(1.5)	(2.4)		(2.0)	(3.0)	(2.5)
Zone IV							
14. Lhuntshi	800	1,600	1,600	—	—	1,000	5,000
	(3.0)	(2.9)	(4.9)			(0.9)	(1.9)
15. Mongar	400	2,900	1,900	—	—	1,700	6,900
	(1.5)	(5.3)	(5.8)			(1.3)	(2.9)
16. Pemagatsel	—	1,600	1,700	—	—	300	3,600
		(2.9)	(5.2)			(0.2)	(1.4)
17. Samdrupj-ongkhar	1,900	4,200	3,600	—	100	400	10,200
	(7.1)	(7.7)	(11.0)		(1.0)	(0.3)	(4.0)
18. Tashigang	2,300	8,600	6,100	—	200	64,200	81,400
	(8.6)	(15.7)	(18.6)		(2.0)	(48.6)	(31.5)
Bhutan	**26,800**	**54,900**	**32,800**	**900**	**9,900**	**132,000**	**257,300**

Source : Statistical Year Book of Bhutan 1990.

** Area for Tsamdok and Sokshing.

Agricultural practices are mostly labour-intensive with little use of modern technologies. Most of the labour is provided by the family and land is "temporarily reassigned between relatives"[6] to keep the ratio of labour to land undisturbed. However, at peak period or during harvest period labour is exchanged against a share of the crop except in the south where landless labourers coming from across the border are employed on a wage basis. Labour shortages, especially in rice cultivation limit the scope of cropping intensities. In order to solve labour scarcity in rice cultivation, the entire process takes place on an extended basis. For instance the crop is established over an extended period — February to July — in dry-bed nurseries, curtailing demand on labour and harvesting is also done on an extended period reflecting variable planting dates, uneven crop maturity and threshing methods often leading to delays in establishing the winter crops, particularly wheat. On the whole, the scope of double cropping becomes seriously restricted, specifically in the south and east due to inadequate irrigation facilities though in the west it has risen to 160 per cent.

The distribution of land under different types of agricultural uses taken for each district of Bhutan gives an interesting picture. The country consists of 18 districts (dzongkhags) and five of them — Samchi, Chirang, Geylegphug, Samdrupjongkhar and Tashigang — lying in the south contain 63.4 per cent of the total wetland cultivation (Table 4.4). Besides, there are at least four interior valley districts — Paro, Thimphu, Wangdiphodrang and Punakha — which together contain 19.4 per cent. Thus, altogether more than 83 per cent of the total wetland cultivation is confined to these nine districts occurring in the south and east, and consisting primarily of flat valley floors. In contrast, wetland cultivation becomes much less significant in the remaining nine districts lying in the western, central and northern parts of the country due to rugged mountainous terrain and there is at least one district in the country — Bumthang — where wetland cultivation seems to be non-existent. At the same time, dryland cultivation the most dominant type among the agricultural uses besides pasture and grazing, having over double land occupance than wetland cultivation, features equally important like the latter, that is wetland cultivation in the same five districts of Tashigang, Samchi, Chirang, Geylegphug and Samdrupjongkhar in order of importance, having altogether 64.5 per cent of the total land under dryland cultivation. But the case, rather, is different with the interior valley districts like Paro, Thimphu, Punakha and Wangdiphodrang where dryland cultivation makes a comparatively lower profile, being altogether absent in Punakha, while in the smaller interior valleys of the districts like Mongar (5.3 %), Bumthang (2.0 %) etc. the dryland cultivation is more important than wetland cultivation.

In comparison, agricultural practices representing orchards and plantations or horticulture in general takes third place in terms of land occupance, covering hardly 4 per cent of the total land under agriculture. Unlike the other two mentioned above, the major part of the horticultural practices is concentrated in three southern districts — Samchi, Chirang and Geylegphug, — using more than 77 per cent of the total land under this practice. It is completely absent in seven districts and besides Chhukha, having over 7 per cent of the total land under horticulture, its place is not a very significant one. Kitchen garden takes a back seat among agricultural practices in terms of land occupance though its importance as almost an inseparable part of farming households cannot be ignored.

In contrast, the place of tseri or shifting cultivation seems to be quite significant in overall land occupance, taking third place, engaging over 12 per cent of the total agricultural land. Barring Punakha district, it is almost ubiquitous in its appearance, occurring in 17 out of a total of 18 districts of Bhutan and though widespread in practice, five districts, namely Tashigang, Chhukha, Samchi, Samdrupjongkhar and Shemgang lying in the southern part of the country contain nearly 67 per cent of the total land under tseri or shifting cultivation.

Agricultural land shown in Table 4.4, under the head of 'Others', includes pastures and grazing land comprising the largest unit, representing 51.3 per cent of the total. Its presence is, perhaps, more ubiquitous than that of land under tseri practices since its overwhelming presence is noted in the sheer volume of land it occupies in many districts — the largest category in at least seven of them — Tashigang, Ha, Punakha, Thimphu, Chhukha and Shemgang.

On the whole, it is the districts in the south that take the leading place in the agricultural activities of the country and in this respect considering all the four principal types of cultivation, such as wet, dry, tseri and orchards, Samchi takes the leading place containing 16.6 per cent of the total cultivated land under the four major heads mentioned above, followed by Tashigang (13.8 per cent), Geylegphug (12.1 %), Chirang (10.6 %) and Samdrupjongkhar (7.9 %). In other words, a total of 61 per cent of the land put under wet, dry, tseri, and orchard cultivation is concentrated in these five districts of Bhutan.

Plate 7 : Motithang Hotel, Thimphu

Plate 8 : Agricultural Terraces on way to Wangdiphodrang

Major and Minor Crops

On the basis of wide diversity in altitude, slope, climate and soil characteristics, Bhutan is divided into five agro-ecological zones. They are : (i) Temperate, (ii) Warm Temperate, (iii) Sub-Tropical, (iv) Humid Sub-Tropical, and (v) Wet Sub-Tropical. A variety of crops are grown in these zones depending on physical conditions and requirements of the inhabitants.

Rice and maize are the principal cereals grown in all the zones, on about 65 per cent of the cultivated land in varying degree of concentration (Table 4.5). As a matter of fact, maize is the most important crop in areal coverage, taking up 40 per cent of the land devoted to the production of all the annual crops in the country, since this is the largest coverage made by any single crop in the whole of Bhutan. In fact, cereals take a very important place in cultivation, with an occupance of over 88 per cent (Table 4.6) of the total land under the cultivation of annual crops. Maize being a hardy crop is cultivated widely but its major concentration is in the sub-tropical zone of the south. Rice is the principal wetland crop grown in the monsoon season on a much wider scale, reaching altitudes upto 2,500 m. In total production, rice is by far the largest crop of the country. Next in importance among the cereals are wheat and buckwheat, mainly concentrated in the temperate zone though wheat is also grown as a winter or dryland crop in the warm temperate and sub-tropical zones where irrigation water is available. However, wheat, the third most important crop, is not preferred as a staple crop. It is grown as a subsistence crop by the farmers and used as subsistence food prepared to take the place of

Table 4.5 : Cropping Intensities by Land Type 1982

(in %)

	Cropping Intensity		
	Wet Land	Dry Land	Tseri Land
West	161	111	19
Centre	131	117	23
East	131	135	20
South	122	110	30
Total	134	116	24

Source : Bhutan, The World Bank, Washington, 1984.

Table 4.6 : Area Under Different Crops (Annual) in Bhutan 1988-89

Crop	Area under the crop (hectares)
1. **Cereals**	
Rice	26,030
Maize	41,890
Wheat	6,420
Buckwheat	7,190
Barley	2,580
Millet	7,570
Other Cereals	1,370
All Cereals	93,060
2. **Legumes**	
Soyabeans	1,960
Other Beans	690
Pulses	1,220
All Legumes	3,870
3. **Oil Seeds (mainly Mustard)**	4,120
4. Roots and Tubers	
Potatoes	2,460
Ginger	360
Radishes	240
Turnips	150
Tapioca	390
All Roots and Tubers	3,720
5. **Vegetables**	630
6. **Other Crops**	
Jute	90
7. **All Other Crops**	130
All Crops (annual)	105,540

Source : Statistical Year Book of Bhutan, 1990.

rice. The yields are generally low as compared to other South Asian countries but its popularity is sure to increase with more urbanisation coupled with a growing demand for bread and related products signs of which are clearly visible in the emergence of confectioners in the urban places of the country. Barley, potato and mustard seeds are primarily the products of the temperate zones and grown with the aid of irrigation in the warm temperate as well as in sub-tropical zones. Rainfed crops and most of the orchards are cultivated on unterraced land. Millet makes another important cereal. Potato has the widest coverage among all the vegetables grown as a temperate and warm temperate crop in the mid-elevation areas. Legumes have nearly 4 per cent of the total cropped area (Table 4.6) and soyabeans have already exceeded the pulses in the amount of land occupied by it. They along with vegetables are grown widely in the country, the variety changing depending on physical conditions and needs of the people. Jute grown only in the south on a very small scale is one of the new entrants.

The geographical distribution of four major crops shows this concentration more clearly. These crops are paddy or rice, maize, wheat and mustard. Paddy being an irrigated crop, is mostly confined to the river valleys where land is terraced for its cultivation. The major concentration of this crop has taken place in five districts of the south lying in the sub-tropical zone: in order of land occupance they are Geylegphug, Samchi, Chirang, Tashigang and Samdrupjongkhar, containing together 16,380 hectares (Table 4.6) or 63 per cent of the total rice land. There are at least three interior districts, namely Punakha, Paro and Wangdiphodrang where rice is an important crop grown on the floors and the adjoining terraced slopes of the wide river valleys. Between them the three districts contain about 17 per cent of the total land under rice. Except Bumthang, it is grown in all the districts of Bhutan in varying amounts.

Though maize may be taken as the largest crop in Bhutan in terms of land occupance, it comes next to rice in total production. In spite of its wide coverage like rice, grown in all the districts with the sole exception of Paro, maize is found to be heavily concentrated in the same five districts of the south like rice. The districts are — Samchi, Geylegphug, Chirang, Tashigang and Samdrupjongkhar which, together, contain 75 per cent of the total maize land of the country while another about 19 per cent is concentrated in four other districts, namely Mongar, Daga, Chhukha and Shemgang lying in the sub-tropical zone of the south.

Wheat, on the other hand, though more ubiquitous in its presence than rice or maize, grown in all the districts of Bhutan, is much less important in areal coverage as well as total production. Apart from Paro, having nearly 20 per cent of the total wheat land and Wangdiphodrang with a little over 12 per cent, it appears to be a widely distributed crop with less concentration in any particular zone.

Mustard, as the major oilseed, shows rather greater concentration in the four districts of the south — Chirang, Geylegphug, Samdrupjongkhar and Tashigang, together having about 58 per cent of the total mustard land of Bhutan. There are at least two districts — Ha and Bumthang — where mustard is absent.

The extreme diversity in physical conditions is at the root of development of rich horticultural practices with a great variety of products many of them having been introduced in recent decades. A substantial amount of dryland, comprising about 8 per cent of the land under cultivated crops, has been turned into orchards in response to higher profits obtained from horticultural crops. On the basis of the production of fruit crops at various altitudes the country may broadly be divided into three zones (Table 4.7).

Table 4.7 : Suitable Altitude for Fruit Production in Bhutan

	Areas	Approximate Range of altitude	Important Fruits that can be successfully grown
1.	Low Hills and Valleys near the plains	365 - 914 m (1,200-3,000 ft)	Sub-tropical fruits like citrus, pine apple, mango, guava, lichi, loquat etc.
2.	Mid-Hills	914 - 1,523 m (3,000 - 5,000 ft)	Stone fruits like peach, plum, apricot, persimmon, pear (hard and soft varieties), pomegranate etc.
3.	High Hills and Interior Valleys	1,523 - 2,743 m (5,000 - 9,000 ft)	Apple, Cherry, Pear (soft variety), Walnut, Chestnut, Almond, Apricot etc.

Source : *Grading and Packing of Apples, Farmers Guide,* Department of Agriculture, Ministry of Development, Royal Government of Bhutan, Edition I, March 1981.

Among the wide range of temperate and sub-tropical fruits produced in the country, oranges, bananas and apples take the leading position in terms of the number of trees of each, no doubt, but oranges and apples have a larger outside market for their high quality in which respect bananas have yet to be improved in quality (Table 4.8).

Apple is the most important temperate fruit grown in all the zones with a greater concentration in the western districts, particularly in Thimphu, Chhukha, Ha, Paro and Samchi. Similarly, the districts of Samchi, Ha, Chhukha and Paro have the largest concentration of orange plants, amounting to over 44 per cent of the total for the country. In contrast, banana is more concentrated in the eastern part, Zone III and IV together having nearly 67 per cent of the total number of plants (Table 4.8). Similarly, many of the sub-tropical fruits like guava, mango, jackfruit, lichi, papaya and persimmons prefer the eastern districts. Walnut is absent in Thimphu, Zone I and Zone II but it is mainly concentrated in Zone IV, representing the districts lying in the east and south, such as Tashigang, Samdrupjongkhar, Mongar, Pemagatsel and Lhuntshi, having the bulk of the plants.

Table 4.8 : Horticultural Crops in Bhutan 1988-89

('000 trees)

Crop	Thimphu	Zone I	Zone II	Zone III	Zone IV	Total
Temperate Fruits						
Apples	49.7	93.6	13.4	47.0	31.8	235.5
Apricots	0.8	1.5	4.9	0.7	3.2	11.1
Peaches	2.0	7.6	29.3	5.9	32.4	77.2
Pears	1.1	2.8	13.0	3.5	7.8	28.2
Plums	0.5	0.8	3.4	0.8	5.6	11.1
Walnut	-	-	-	1.9	17.4	19.3
Mangosteen*	-	-	-	-	2.4	2.4
Sub-Tropical Fruits						
Oranges	4.6	1,148.3	504.8	762.1	175.4	2,595.2
Lemons	0.3	74.5	9.1	23.5	3.1	110.5
Bananas	1.1	86.7	182.8	286.6	258.2	815.4
Guavas	0.7	9.3	25.2	10.4	15.6	61.2
Mangoes	-	3.3	2.3	9.9	3.2	18.7
Jackfruit	-	3.1	3.0	4.0	2.2	12.3
Lichi	-	-	-	3.8	0.7	4.5
Papaya	-	-	-	3.1	2.1	5.2
Persimmons	-	-	-	-	1.9	1.9
Other Tree Crops						
Cardamom	-	5,610.4	5,318.3	4,054.5	20.5	15,003.7
Arecanut	-	262.9	-	347.3	34.6	644.8
Tea	-	8.0	2.8	5.3	0.5	16.6

Source : Statistical Year Book of Bhutan 1990.

* Mangosteen is considered by many as the world's best flavoured fruit, a dark red-purple berry 2 to 3 inches in diameter having 5 to 7 fleshy, white segments which are eaten. 'The texture of the pulp of these segments is so delicate that it melts in the mouth with a delicious flavour' — *Encyclopedia Americana,* New York, 1973, Vol.18.

There are at least three commercial tree crops of great market value. They are cardamom, arecanut and tea, the last one being newly introduced. The number of cardamom bushes amounting to 15,003,000 is significant considering the amount of fruit they produce. They are mostly concentrated in the districts lying in the western, southern and central parts of Bhutan with over 99.8 per cent of the total number of cardamom bushes. Similarly, the cultivation of tea has so far remained confined to these three zones — I, II and III — where 97 per cent of the tea bushes are concentrated.

The yield rate of crops vary widely depending primarily on physical conditions though its enhancement is done through such inputs like irrigation, high yielding variety of seeds, chemical fertilizers and green manures. Rice shows a wide variation, from a minimum of 880 kg per hectare in Ha district in the extreme west to a maximum of 2,670 kg in Punakha in the north when the average for the country is 1,660 kg (Table 4.9). The variation in yield, however, is not that wide for wheat. There are two districts — Lhuntshi (1,090 kg) and Geylegphug (1,030 kg) — situated in the north-east and south respectively where the yield rate of wheat is comparatively much higher than the national average (640 kg) per hectare. In other districts the rate of yield more or less veers round the national average for wheat.

Maize, however, gives a different picture, districts lying in the eastern and southern parts providing a higher yield which is more than double in some districts than the national average (740 kg) per hectare. For instance, Lhuntshi (2,170 kg), Mongar (1,760 kg) and Tashigang (1,590 kg) maintain a higher yield while in three other districts — Shemgang (1,180 kg), Bumthang (1,160 kg) and Pemagatsel (1,060 kg) — the yield rate is considerably higher than the national average. With the exception of these districts and another, Punakha (770 kg), the yield rate of maize falls below the national average.

On the contrary, mustard does not exhibit this wide variation in its yield rate, which remains near to the national average in general. Nevertheless, districts in the east and south maintain a higher rate of yield, the largest (600 kg/ha) being in Pemagatsel while in seven districts it is lower than the national average (Table 4.9).

Table 4.9 : Area, Production and Yield of Major Crops in Bhutan 1988-89

District or Dzongkhag	Paddy			Wheat			Maize			Mustard		
	Area (ha)	Production (tonnes)	Yield kg/ha	Area (ha)	Production (tonnes)	Yield kg/ha	Area (ha)	Production (tonnes)	Yield kg/ha	Area (ha)	Production (tonnes)	Yield kg/ha
Thimphu	760	1,850	2,420	400	260	660	30	10	500	70	20	300
Chhukha	920	950	1,030	310	220	700	1,830	680	370	300	80	270
Ha	90	80	880	460	230	510	70	40	540	—	—	—
Paro	1,460	2,400	1,640	1,250	650	520	—	—	—	120	50	390
Samchi	4,400	5,740	1,310	250	130	500	7,540	2,490	330	230	80	340
Chirang	3,050	4,510	1,480	340	170	490	6,380	2,910	460	610	170	270
Daga	790	1,070	1,360	80	40	500	2,360	890	380	320	100	310
Punakha	1,600	4,290	2,670	420	230	560	80	60	770	160	50	310
Wangdi-phodrang	1,250	na	na	790	na	na	60	na	na	110	na	na
Bumthang	—	—	—	260	150	570	10	20	1,160	—	—	—
Geylegphug	4,690	7,120	1,520	390	400	1,030	6,990	2,650	380	560	160	280
Shemgang	770	900	1,170	200	150	780	1,060	1,250	1,180	170	50	290
Tongsa	600	820	1,370	270	180	640	380	270	720	40	10	390
Lhuntshi	800	1,640	2,050	150	170	1,090	1,140	2,480	2,170	70	30	440
Mongar	470	470	980	130	60	470	2,570	4,530	1,760	80	30	400
Pemagatsel	120	110	960	160	100	630	860	920	1,060	30	20	600
Samdrup-jongkhar	2,030	2,910	1,430	60	30	490	4,700	2,650	560	690	180	260
Tashigang	2,210	4,930	2,230	490	460	940	5,810	9,230	1,590	500	280	550
Bhutan	26,030	43,140	1,660	6,420	4,080	640	41,890	31,130	740	4,080	1,340	330

Source : Statistical Year Book of Bhutan, 1990.

Agricultural Landholders

According to the Central Statistical Office of Bhutan, there were 63,480[7] landholders in the country, majority of them (56 %) concentrated in the five southern districts — Tashigang, Samchi, Chirang, Geylegphug and Samdrupjongkhar — of which Tashigang alone had nearly 20 per cent.

The majority of the landholders have a small amount of land which hardly gives an economic return, particularly where terrain is rough. It may be observed that 27.7 per cent of the farmers have less than 1 hectare of land and another 30.6 per cent have less than 2 hectares per family. For about 18 per cent, it is more than 2 but less than 3 hectares per household while it does not exceed 3.00 — 4.99 hectares for about 14 per cent of the landholders (Table 4.10). A little less than 10 per cent of the total farming households in the country possess over 5 hectares each and the largest number of them (17.2 %) belong to Samchi district followed by Tashigang (15.8 %), Chhukha (12.8 %) and Geylegphug (11.9 %). Thus, it appears that the larger farmers are more concentrated in the southern districts than elsewhere. This is also true about farmers having landholdings amounting

Table 4.10 : Agricultural Landholdings by the Size of Land in Bhutan, 1988-89

	Size of Landholdings (hectares)						
	0.01- 0.49	0.50- 0.99	1.00- 1.49	1.50- 1.99	2.00- 2.99	3.00- 4.99	5.00 and over
Bhutan	5,900 (9.4%)	11,530 (18.3%)	11,270 (17.9%)	7,990 (12.7%)	11,420 (18.1%)	8,690 (13.8%)	6,150 (9.8%)

Source : Statistical Year Book of Bhutan, 1990.

to 3.0 to 4.99 ha. For instance, Samchi tops the list with 18.3 per cent of the total land holdings of this size-group followed by Chirang, having 13.6 per cent, Geylegphug 11.6 per cent, Tashigang 11.4 per cent and Samdrupjongkhar 9.4 per cent, which is much lower in other districts.

Some Problematic Issues in the Agricultural Sector

It may be worthwhile to note that 87.2 per cent (1984) of the population of Bhutan between age-group 15 — 64 years is dependent on agricultural pursuits for their living and this exceeds 90 per cent for the people belonging to age group 35 — 64 years.[8]

In a situation like this agriculture gets top priority for development which formed a basic policy since the beginning of the Five Year Plans in 1960. There are, however, certain problems plaguing agriculture for a long time, stalling its normal progress. One of them is tseri cultivation which being an environmentally degrading practice, badly affects the economy as well as the carrying capacity of the land where this is practised. As mentioned earlier, engaging about 26.2 per cent of the cultivated land, it is a phenomenon more of the east and south. According to an estimate, in the east 18,461 households are dependent on tseri culture, burning down a minimun of 30-40 acres to cultivate just 10 acres for livelihood.[9] This means colossal wastage of the most important resource of the country — arable land—amounting hardly to 8 per cent of the total land area of the country. The planners seem to be in favour of improving the existing situation by way of offering monetary incentives already introduced for converting tseri land into dry or wet cultivation or into orchards. There is, however, a completely opposite view regarding tseri cultivation according to which 'tseri is sustainable and does not cause degradation if there is a fallow period of at least 3 years, and the slope is less than 8 per cent'.[10]

Of late, wild boar has turned out to be one of the worst menaces as expressed in an *Agricultural Newsletter* : "There is a new competitor in the farms for the Bhutanese farmers — the boar".[11] A survey conducted by the Ministry of Agriculture in 1993 revealed that in 12 out of 13 districts covered, crop losses due to wild animals — mainly boars — amounted to 17.5 per cent and recorded as high as 29 per cent in Pemagatsel and 28 per cent in Tongsa.[12] According to the report in the *Agricultural Newsletter* (No. 44, January-February 1995), the wild boar causes extensive damage to the crop consuming a wide range of seeds, fruits, young leaves, tubers and succulent stems, reptiles and insects besides causing damage by wallowing and trampling. Damages affecting 5 to 50 per cent of the cropped area and the loss of yield going from 10 to 100 per cent in certain cases have been reported.[13] According to farmers, the rapid increase in boar population in recent years was due to the reduction of its main predator, the wolf resulting from wanton killing, controlled forest fires, social taboos and forest laws.

Like wild boars, pests like porcupine and locusts have been reported to do serious damages to the crops. Citrus Fruit Fly is another major pest in Bhutan, causing heavy damages to oranges, sometimes destroying over two-thirds of the crop.[14] Another problem is apple die-back detected first in 1990 which stunts the growth of the apple plants resulting from boron deficiency. It has been found through experiments that application of borax in the fertilizer used for apple plants will solve the problem.

Developmental Strategies for Agriculture

It is reported in one of the issues of the *Agricultural Newsletter* in 1993 that the Government of Bhutan, in collaboration with the F.A.O. initiated a comprehensive programme to ensure food security for the people so that adequate food is available for an 'active and healthy life'.[15] It is, however, maintained that food scarcity in Bhutan is not as acute as it is in many other countries but it is necessary to ensure growth in food production to match with the growth of population, its projected increase of 2.5 per cent per year.

In the absence of accurate data, the total population of Bhutan was estimated to be around 1.165 million in 1982, with an average density of about 25 per sq mile and a growth rate in the range of 1.8 per cent to 2.0 per cent. At that time, 95 per cent of the working population was engaged in agriculture and allied activities, comprising primarily subsistence farming and animal husbandry. Although Bhutan had been self-sufficient in food for a long time, food production could not keep pace with the rate of population growth in the recent decades. On the other hand, agriculture suffering from a number of constraints like limited irrigation facilities, lack of quality seeds, fertilizers, rural credit, proper transport, storage and marketing opportunities, could not prosper as desired. At the same time, increase in the rate of consumption , particularly in urban areas as a result of rise in living standard in a growing economy and emergence of non-farming communities added further to the problem.

Agricultural development has enjoyed a high priority in all the development plans. In the First Five Year Plan (1961-66), road development got top priority receiving two-thirds of the total financial allocations followed by manpower development and education. Road development continued to receive same priority in the Second Five Year Plan (1966-71) though this time with one-third of the total plan expenditure with commensurate increase in education and for the first time agriculture allotted with 10 per cent became a priority sector. In the Third Plan (1971-76), the pattern remained the same but allocation for road development was reduced and in the Fourth Plan (1976-81), the major shift in emphasis took place, giving top priority to the development of agriculture attaching greater importance to animal husbandry, natural resources and forestry.[16] Thus, while agriculture was allocated with a bare 1.8 per cent (1.9 Nu million) of the total outlay for the First Five Year Plan (1961-66), it gradually increased from 10.7 per cent (21.6 Nu.m.), in the Second Plan (1966-71) to 12.3 per cent (55.3 Nu. m.), in the Third Plan (1971-76) and 23.5 per cent (259.0 Nu.m.), in the Fourth Plan (1976-81).[17] By the end of the Fourth Plan, about 90 per cent of the total population remained dependent on agriculture which, along with animal husbandry, generated 47 per cent of the GDP.[18] Development activities concerning agriculture during different plan periods may be summed up under two specific categories as follows : (a) establishment of agricultural support services and institutions; (b) land development and irrigation schemes. The establishment of extension services with trained personnel receiving particular emphasis from the First Plan had been completed in all the 18 districts of Bhutan by the end of the Fourth Plan along with the establishment of four training centres for extension workers. The extension services included improved seeds and implements, fertilizers, fruit seedlings etc. at subsidised rates for the farmers. For this purpose a countrywide network of agricultural farms were set up for the multiplication of seeds and seedlings, equipped with facilities for giving demonstrations to the farmers about the proper use of equipments and new technologies. Further, specialised farms were established for research and development and the propagation of important commercial crops like potato, apple, oranges, cardamom, etc.[19]

The impact of all these developmental prograrnmes was directly reflected in improvement clearly noticeable in this sector by the end of the Third Plan (1971-76) as recorded in increases in the area under cultivation and area with irrigational facilities. The gross cropped area under cereal production increased from about 200,000 acres in 1973 to about 240,000 acres in 1978 and 278,000 acres in 1982.[20] Cropping intensity depending on population pressure and the availability of irrigation water ranged from 122 per cent in the south to 161 per cent in the west for wet cultivation and from 110 per cent in the south to 135 per cent in the east for dry cultivation (Table 4.5). In spite of that 25,000 M.T. of cereals had to be imported from India to meet the deficit by the end of the Fourth Plan.

In the Fifth Plan (1981-87) agriculture accounted for 11.5 per cent of the total allocation preceded by Industry and Mines (17.4 %), Power (16.5 %) and Public Works Department, that is, roads, bridges, water supply and urban development (12.4 %). In priority, agriculture seems to have scaled down to the fourth place when greater emphasis was laid on 'revenue generating sectors and ... power'. However, in terms of actual outlay, the amount was nearly double, increasing from 259.0 Nu.m. to 494.8 Nu.m. for agriculture.

While the major efforts during the four previous plans were directed towards the creation of a comprehensive infrastructure and the opportunity for a vigorous thrust on agricultural development, the Fifth Plan clearly outlined the principal objectives, putting greater pressure on (a) increasing farmer's income through the diversification of production, mainly cash crops, (b) increasing improved seeds, production, and achieving self-sufficiency to the extent possible, (c) improving the overall nutritional status of the population and (d) achieving self-sufficiency in the production of foodgrains.

By this time limits to any more increases in land to be brought under cultivation having been reached, and there being labour shortages ultimately changed the focus to increasing the intensity of cultivation during the Fifth Plan period, and future production increases were to come from increased labour productivity resulting from better quality inputs — that is, greater use of improved seeds, fertilizers, irrigation and improved farming methods.

Likewise, the Fifth Plan laid emphasis on programmes for the multiplication and the supply of improved seeds and seedlings to be undertaken in all the districts of the country with a target of 5,200 metric tonnes of improved seeds and 4.1 million fruit seedlings and rhizomes to be distributed among the growers. In order to improve soil quality, construction of compost pits were to be encouraged through subsidies and all extension centres were to supply fertilizers as well as provide guidance on their use. The transport cost for fertilizer was to be subsidised by the Government. The target for chemical fertilizers and micronutrients was set at 2,810 metric tonnes and the number of compost sheds at 127,000.[21]

In order to overcome the labour shortages turning out to be a major impediment to increasing crop intensity, particularly at the time of transplanting/sowing and harvesting of cereal crops, introduction of simple tools and manual/power-operated equipment to be sold at reasonable prices to the farmers was to be taken up along with training facilities. For this purpose, the Plan targeted to procure for sale or hire 35 tractors, 112 power tillers, 1,261 manual and power-operated machinery and 612 farmers were to be trained during the Plan period.

Plant protection, rural credit, improvement of extension centres, programme for importing scientific knowledge etc. featured in the Plan for agricultural development. The importance of growing vegetables on a large scale was highlighted and the National Potato Development programme aimed at greater improvement in quality and yield, increase in the production and distribution of quality seeds and reduction in the incidence of diseases afflicting potato crop. Food processing based on surplus crops was to get a further impetus through setting up new plants at Tashigang, Lhuntshi, Mongar and Pemagatsel in addition to those two working at Paro and Bumthang.

Three major irrigation systems were to be constructed during this period in Geylegphug-Bhur area of south-central Bhutan consisting of three projects : (i) Bhur Project, designed to be a gravity-fed system with a command area of 800 ha benefiting 600 farmers; (ii) Taklai Project, another gravity-fed system with a command area of 2,400 ha; and (iii) Geylegphug Area Development Project Phase II, a lift irrigation system, having a command area of 1,200 ha. These, along with minor irrigation projects will add altogether 13,100 ha to the net command area under irrigation in the country.

Conversion of dry land into wet land with the aid of irrigation and land under tseri or shifting cultivation into permanent cultivation or dry land practices were going to be much encouraged involving financial subsidy . Targets laid down for the Fifth Plan in this respect were as follows :

8,800 ha of dry land to be converted into wet land,
8,800 ha of tseri land to be converted into permanent fields,
2,250 ha of land to be contour-bunded, and
40 ha of land drainage work to be undertaken.[22]

Though objectives of the Sixth Plan (1987-92) remained the same as in the previous Plan, agriculture received the largest of the outlays with 17.5 per cent of the total allocation and in terms of GDP, the position of agricultural sector including livestock, forestry and fishing seems to have made a significant progress, providing 46 per cent for the country in which the contribution of agriculture alone was exactly one-fourth or 25 per cent of the total in 1989 (Table 4.11).

Table 4.11 : Gross Domestic Product of Bhutan, 1980-89
(Nu. Million, 1980 Constant Prices)

Heads	1980	1981	1982	1983	1984	1985	1986	1987	1988	1989
Agriculture, Livestock, Forestry and Fishing	621.4	636.1	692.2	742.2	806.5	833.9	881.0	925.8	939.7	962.9
Agriculture Proper	309.9	312.5	331.0	368.9	388.3	411.6	436.3	458.1	485.6	524.7
Livestock	139.2	144.5	149.7	155.4	161.3	168.6	177.4	184.5	191.5	201.7
Forestry & Logging	172.3	179.1	211.5	217.9	256.9	253.7	267.3	283.2	262.6	246.5
Mining & Quarry	6.8	8.9	12.1	8.9	15.8	12.6	22.2	21.6	19.0	21.7
Manufacturing	35.8	59.1	59.6	62.9	67.2	75.4	71.0	105.0	110.3	128.2
Electricity & Gas	2.5	2.7	3.1	6.0	5.2	6.0	60.4	229.0	225.1	231.5
Construction	88.5	131.0	157.7	185.2	173.5	169.0	141.8	152.3	129.0	133.8
Wholesale & Retail Trade, Restaurants and Hotels	121.5	140.3	132.0	122.0	123.4	132.4	143.4	142.4	129.2	133.6
Transport, Storage & Communications	47.9	54.9	57.4	64.8	66.4	79.4	83.9	91.3	122.1	142.4
Financing, Insurance, Real Estate & Business Services	70.2	80.3	77.0	91.1	109.4	110.1	126.2	135.7	141.0	163.5
Community, Social & Personal Services (Government, Administration and Defence)	120.4	121.5	102.1	113.1	130.3	126.0	168.6	200.0	210.0	217.0
Less : Imputed Bank Services	-20.0	-30.0	-24.0	-26.0	-32.0	-25.0	-24.0	-30.0	-31.8	-41.1
Gross Domestic Product (GDP)	1095.0	1204.8	1269.2	1370.2	1465.7	1519.8	1674.5	1973.1	1993.6	2093.5

Source : Seventh Five Year Plan, 1992-97, Bhutan.

In fact, agriculture has maintained this lead since 1980, for which data is available, even though its share has gradually shrunk from 57 per cent of the GDP in 1980 to 46 per cent in 1989 when contribution from other sectors, like power (electricity and gas), manufacturing and transport, storage and communications made a significant progress.

In agriculture, the Seventh Plan (1992-97)[23] stressed on the same objectives, such as : (a) to increase self-sufficiency in staple food; (b) to increase the per capita income of the farmer, and (c) to increase the productivity per unit of farm labour and agricultural land. The strategies to be adopted for realising those objectives were laid down involving the improvement of the provisions of services and incentives to the farm population with more effective steps for reaching them through Area Development Programmes.

The Area Development Programmes implemented during the Sixth Plan period took an important role in increasing the use of fertilizers and improved seeds. However, the idea came under heavy criticism as the service offered by the ADP comprising high level technical support, equipment and other facilities covered only a small area when the remaining areas of the country had to accept the reality of the situation with fewer services, leading ultimately to great spatial disparities. Besides, financially as well as technically it was beyond any means to maintain such high levels of assistance for a long time which further necessitated a thorough review of the matter.

During the Sixth Plan period new high yielding varieties of rice were developed and distributed among the farmers and particularly those in ADP areas started using these varieties. New maize, wheat and oilseed varieties were ready for use by the end of the Sixth Plan. Green manuring adaptive research showed promise for increasing soil fertility. Further, research was being carried out specifically on mushroom propagation and the production of essential oil as the cooking medium, both having great commercial potential.

The 'Research and Extension Division' came up for coordinating the activities of the Extension Centres in offering services to the farmers and it arranged for in-service training for the district's staff. An impressive headway was made during the Sixth Plan in making high-quality inputs. The Ministry of Agriculture which is responsible for the supply of agricultural inputs to the farmers distributed the following items in 1989-90:

Fertilizers — 1,669 metric tonnes
Oilseeds — 11 metric tonnes
Cereal Seeds — 492 metric tonnes
Vegetable Seeds — 1,409 metric tonnes
Temperate Fruit Plants (Seedlings) — 53,012
Tropical Fruit Plants (Seedlings) — 28,200.

However, lack of proper storage facilities, transport and access roads were reported to be the major handicaps in the supply of inputs. Besides, soil conservation programmes including contour bunding, terracing and conversion of tseri land into permanently cultivated land had also been taken up in all the districts. New techniques for the conservation of soil, like ditch and bund, vegetable bunding, and strip cropping were experimented in Chirang and Punakha districts where watershed management programmes were also introduced. Lastly, the marketing system went through drastic changes during this period with complete privatisation of the Food Corporation of Bhutan's retail operations. The marketing of orange and apples in Bangladesh, beginning at this time, indicated great potential of horticulture as an earner of foreign exchange.

Seventh Plan Objectives for Agricultural Development and Future Prospects

The Seventh Plan objectives were nothing but a continuation of those adopted in earlier Plans, aimed primarily at (a) attaining a suatainable development of agricultural production to make the country self-reliant for food requirements; (b) improvement in income, living and nutritional standards of the rural population, and (c) sustainable utilisation of natural resources.

The growing population could not be matched with a commensurate increase in food production, eventually resulting in imports of cereals and other food items in large amounts which made a heavy pressure on finance (Table 4.12). Rice is the principal item on the list of imports enjoying the pride of place in all the nine years, 1981-82 to 1989-90, and wheat is the other cereal taking an important place. As reported in the Seventh Plan Main Document, "increasing urbanisation, combined with limited arable land and low yields" seem to have made an unholy alliance in bringing

Table 4.12 : Imported Food Items in Bhutan, 1981-82 to 1989-90

(tonnes)

Commodity	1981-82	1982-83	1983-84	1984-85	1985-86	1986-87	1987-88	1988-89	1989-90
Rice	2,345	5,843	6,033	5,388	7,211	11,282	16,750	11,600	9,900
Wheat	873	2,725	2,123	2,599	2,450	3,350	5,767	4,990	2,834
Sugar	1,276	1,381	1,800	1,824	2,749	3,574	4,039	3,616	3,040
Salt	3,475	2,755	2,326	3,598	5,689	3,516	5,035	6,495	4,619
Oil		-	240	306	557	874	1,280	400	340

Source : *Statistical Year Book of Bhutan*, 1990.

the situation to this critical level though signs of improvement have started to appear from the end of the Sixth Plan with visible decline in the quantum of imports. This has happened in the case of both rice and wheat, dwindling in amount from 1988-89 onwards. Sugar as well as oil too registered similar decline from the same year barring salt alone with reduction only in the last year. In other words, implementation of different improvemental steps during the Sixth Plan enhancing agricultural production had a direct impact on bringing down the imports of food items. The planners, however, have realised the fact that self-reliance in cereal requirements appears to be yet a far cry though the great potential in horticultural development and a burgeoning foreign market may provide necessary foreign exchange to buy cereals from foreign markets in future. At the same time, greater emphasis on horticulture providing products of higher unit-value will act as an encouragement for the farmers to produce more and earn more to raise the standard of living. Perhaps, this is a more sure and better way for attaining the objective of the upliftment of the 'living and nutrition standard of the rural population'.

From the point of view of agriculture, sustainable utilisation of natural resources implies the maintenance of soil quality or checking the depletion of soil fertility so that agricultural production can be increased unimpaired. It was further observed in the Main Document of the Seventh Plan that, since a limit has already been reached in the supply of arable land except in cases where previously unused land may be developed, any sustainable development depends largely on conservation practices which demand a close integration between such practices like soil conservation, afforestation, agro-forestry, organic fertilizers, crop rotation etc. It was proposed that the Ministry of agriculture henceforth will assist the farmers to integrate new techniques into the existing production system. Agro-forestry as a new production system is thought to be easily adaptable for farmers for integration into the existing system and they provide the only means for agricultural expansion as an 'environmentally viable alternative to logging' in areas identified for clearing forests.

The implementation of new strategies for increase in production in the arable sector will be done through two programmes adopted for this purpose — the Accelerated Food Production Programme (AFPP) and the Integrated Horticultural Development Programme (IHDP) of which the former will involve increase in cereal production with the introduction of new cropping patterns to improve soil fertility, such as legumes. A

Horticultural Section established under the IHDP will provide assistance to the producers through research, extension services and marketing support. The Potato Development Programme is also going to be an integral part of the IHDP. Irrigation, mechanisation and post harvest losses will form areas of further research and among them emphasis on irrigation development during the Seventh Plan period will primarily centre round the 'rehabilitation of existing schemes, examination of new materials and technologies, exploration of groundwater potential and training of farmers'[25] in management skills through the formation of Water Users Association.

Although irrigation systems have been developed as long ago as about thirty years or so in Bhutan, it is very difficult to maintain them. Table 4.13 gives an idea about irrigation facilities developed for the first time and those renovated, all of them having rivers as their source. Taken

Table 4.13 : Irrigation Schemes completed during 1989-90 in Bhutan

District or Dzongkhag	New Construction			Renovation		
	No. of Schemes	Length of Channels (km)	Command Area (ha)	No. of Schemes	Length of Channels (km)	Command Area (ha)
1. Thimphu	—	—	—	3	6.5	323.9
2. Chhukha	1	2.9	164.0	1	3.5	121.5
3. Ha	2	2.9	143.7	1	1.0	11.9
4. Paro	1	4.3	56.7	4	12.7	113.1
5. Samchi	—	—	—	4	9.4	310.3
6. Chirang	—	—	—	2	3.2	72.9
7. Daga	—	—	—	2	2.9	59.7
8. Punakha	—	—	—	2	4.1	85.4
9. Wangdiphodrang	2	15.5	291.5	4	13.4	233.7
10. Bumthang	—	—	—	—	—	—
11. Sarbhang	—	—	—	3	6.0	91.9
12. Shemgang	—	—	—	1	1.8	7.3
13. Tongsa	—	—	—	2	5.6	32.4
14. Lhuntshi	—	—	—	1	2.7	121.5
15. Mongar	—	—	—	—	—	—
16. Pemagatsel	—	—	—	1	2.0	34.4
17. Samdrupjongkhar	1	1.4	68.0	2	6.9	487.9
18. Tashigang	—	—	—	1	1.2	54.3
BHUTAN	7	27.0	723.9	34	83.0	2,162.1

Source : Statistical Year Book of Bhutan, 1990 .

together, irrigated land under these two heads represents 1.1 per cent of the total land under agriculture, and if one excludes the land used for grazing etc., the irrigated land comprises 2.3 per cent of the total cropped land of the country. The construction of new channels for a length of 27 km during the Sixth Plan is quite an impressive progress made in this regard in a mountainous country under great many constraints, and renovation of the existing channels for a length of 83 km is another commendable feat. Two districts — Samdrupjongkhar and Wangdiphodrang — contain the major part of the irrigation channels, having about 20 and 18 per cent respectively.

The Agricultural Machinery Centre (AMC) at Paro was founded in 1983 and it has supplied 75 power tillers in the first part of the Seventh Plan period, nearly 20,000 tools and implements, 275 threshers 41 power chain saws and 4,000 metres of gardening pipes.[26] The Regional Protection Centre to be established at Tashigang and Geylegphug, will conduct research for providing services for plant protection and supply of chemicals for the purpose was to continue. As reported in *Agricultural Newsletter*[27] — use of pesticides has increased from early 1940s to late 1960s, and till early 1980s, more than 30 types of pesticides were being used in the country, including highly toxic and persistent chemicals like Thimet and Aldrin banned or restricted in most developed countries. From 1989 onwards, the list of pesticides has been reduced to almost half and highly toxic and persistent ones have been phased out. One of the harmful practices mentioned here providing a glaring example of misuse of pesticides out of sheer ignorance, was the regular spraying of apple orchards in Thimphu, termed as 'calendar spray', causing a serious side-effect with an outbreak of red spider mites in apples which has appeared in most apple orchards in Paro and Ha as well. The attitude of the National Plant Protection Project is completely different in this regard putting greater emphasis on an integrated pest management approach which includes biological control (parasites, predators), mechanical control (hand picking etc.), cultural control (age-old practices), genetic control (plant resistance) and chemical control, giving more emphasis on non-chemical ways of pest control as the best long term strategy. In other words, it means a combination of management practices with chemicals rather than an indiscriminate use of the latter.

A Wide Horizon for New Horticultural Crops

The performance in the agricultural sector indicates growing popularity of certain new crops so far as the outside demand is considered which makes an important issue in a country in great need of foreign exchange to boost its economic development.

Among the new crops, apples and oranges occupy premier position. Apple orchards started to be established from early 1962 and by now the farmers growing apples have become conscious about its value in the foreign market. Cultivated in temperate areas, already more than 1,173 hectares (2,900 acres) of land in Thimphu and Paro have been converted into apple orchards and the products of high quality find market in India and Bangladesh. Orange is another crop having wide access to Indian and Bangladeshi markets. Grown profusely in the south below an altitude of 1,500 m and an annual amount of rainfall often exceeding 1,500 mm, the south represents over 90 per cent of the total citrus fruit production in the country, the annual average production being about 23,000 metric tonnes.[28]

Bhutanese farmers have found a new source in walnut for high returns. Walnut is not unknown in Bhutan, generally found in the forests and grown by many one or two in their own garden for a long time. Its commercial importance as a high value crop has however been a recent realisation, encouraging the plantation of exotic varieties and this is found mostly among the progressive farmers in the west of Bhutan where it is taken as an alternative to apples. It is a temperate fruit grown at altitudes 1,500 to 3,000 m, still now mostly for home consumption. However, the planting has increased quite fast in recent years, becoming second numerous to apples in Paro, Thimphu and Ha containing 9,654 walnut trees. A recent survey shows that walnut is also one of the major tree crops in other districts of the country. The production of walnut in Bhutan is likely to be about 100 tonnes by 2000.[29]

Strawberry makes another highly potential commercial crop of Bhutan and although it is not a new crop in Bhutan, growing wild in many areas, farmers have developed interest in growing strawberries adaptable to different climates, ranging from sub-tropical to temperate, which can be grown in open fields as well as under cover, in plastic tunnels and also in greenhouse. The wide varieties make its production possible almost all through the year in different climatic conditions but it is suggested that the crop should be rotated and not preferably be grown for more than three years on the same plot as it is very sensitive to soil-borne pests and diseases.

Asparagus, not a fruit, but a vegetable crop already having a high market demand, is a perennial crop of which the shoots are edible. It is another high value crop with an economic life span of 10 to 12 years, grown under a wide range of climatic conditions, sub-tropical to temperate, at mid to high altitudes upto 3,000 m. It is a very labour-intensive crop usually making a great demand for soil management.[30]

Perhaps, the most lively feature of Bhutanese agriculture is the introduction of new crops, quite often of exotic variety from time to time along with the improvement of the indigenous ones through further research having the major thrust on food crops of high yielding variety adding gradually a competitive character to it. In this regard, research activities started in the eighties and from 1982, efforts are afoot to identify and introduce high yielding varieties of food crops. In 1988, the Variety Release Committee (VRC) was constituted by the Ministry of Agriculture entrusted with the responsibility of helping research in introducing new varieties of crops, their development, evaluation and finally the release of certified varieties ensuring the use of seeds and seedlings by the farmers. The Committee is also empowered to decide about the discontinuation of use of any particular variety of seed or seedling under compelling circumstances. "The ultimate aim of the formation of this committee was to increase crop yield and improvement in the productivity of the land ensuring the food self-sufficiency and economic well-being of the farmers".[31]

The VRC in a meeting held in 1992 released 'Barket' variety as the first high-yielding seed for double rice cropping. In 1994, it released several varieties of crops, each given a local name in recognition of its origin. The type of crops for which different varieties of seeds developed and approved by the VRC for release so that the farmers can use them, appear in the Table 4.14 which shows that two types of mustard seeds, BSA and PT 30 have been released in the name of Bajo Peka 1 and Bajo Peka 2 respectively. This applies to many others though in certain cases the name remains unchanged.

It appears from the Table that eight new improved varieties of seeds were approved for vegetables while another six types of improved seeds for vegetables though not formally approved for release, were already in use as the farmers found them eminently suitable for their high quality.

The official release was made later on. Among the twelve better varictics of apple seeds, six representing the commonly grown ones were approved for release while for six others representing Japanese varieties, provisional approval for the continuation of propagation work was given withholding the final release for the distribution of their seeds to the farmers till adequate information about them were available. The same has also been the case with sixteen varieties of new seeds developed for cherry, plum, and peach the propagation work of which was to continue deferring the release in the absence of necessary information.

The most significant as well as highly encouraging aspect revealed by the long list of the improved variety of seeds of a number of different types of the same crop as well for further multiplication is a clear indication of the undeterred progress of agriculture towards achieving the country's two important goals of self reliance in food as far as practicable, and economic upliftment of the rural community.

Nevertheless, all these depend not on the development and use of improved variety of seeds alone but also on some other controlling factors like the agricultural infrastructure needed to be properly developed to meet the challenging situation in addition to the fulfilment of the requirements in certain other areas as expressed by the farmers during interviews taken on various occasions by the personnel deputed by the government. The complaints primarily concern crop losses due to the depredations of the wild animals, pests and insects, weed control in paddy, soil erosion and low fertility, and the post-harvest losses of the stored grains.

In the opinion of some experts in the field, the existing nature of land use is far from 'optimum' and if not checked or planned in the right manner, it will lead further to the deterioration of the soil quality reducing production potential. Besides, in view of the small size of the farm units, likely to become smaller in future with more fragmentation, a greater thrust on intensive use of land with more emphasis on the production of food crops should get priority (Dorji,1994).[32] Keeping this in view necessary measures will be taken during the period of the Seventh Plan to revert the marginal cropped land to forest and reclamation of suitable forest land for cropping or livestock rearing. At the same time, the RNR (Renewable National Resources) sector, the largest and most important sector of the Bhutanese economy, in discussing the formulation of the Eighth Plan in a meeting held in February 1995, stressed on the following points :

Table 4.14 :Varieties of Seeds released by the Variety Release Committee in Bhutan, 1994

Sr. No.	Crops	Variety	Variety Released As	Remarks
1	2	3	4	5
1.	Mustard	BSA	Bajo Peka 1	
		PT 30	Bajo Peka 2	
2.	Wheat	BL 1093	Bajo Ka-2	
3.	Vegetables			
	Pea	JI 1050	Released as same	
		Snow Pea	— do —	
	Bean	Rajma	Released as same.	
		Long John	do	
	Cabbage	Baldura	do	
		Pride of India	do	
	Okra	Blue Bell	do	
		Pusa Sawani	do	
4.	Backlog of Vegetable Varieties			
	Cauliflower	Snowball-16	do	These seeds developed earlier were already in use but were approved later on as their demand was high among the farmers
	Broccoli	Desico/Decicco	do	
	Onion	Senshu Red	do	
	Lettuce	Sunny	do	
	Summer Squash	Zucchini (green)	do	
	Soya bean	One daughter	do	
5.	Apple	Scion Varieties	Red Delicious	
			Royal Delicious	
			Golden Delicious	
			Jonathan	
			Rich Red	
		Root Stock	MM 106	Provisional Approval given for propagation work
		Nebuta		
		Kogetsu		
		Starking		
		Sekaiichii		
		Jonagold		
		Fuji		

Table 4.14 *(Contd.)*

1	2	3	4	5
6.	Temperate Stone Fruits			
	Cherry	Scion Varieties	Senaka	Provisional approval given for propagation work
			Jabouly	
			Royal Ann	
		Root Stock	Colt	
	Plum	Scion Varieties	Stanely	
			Stantarosa	
		Root Stock	Saint Julian	
	Pear	Scion Varieties	Kosui	
			William Bartlet	
			Hosui	
		Root Stock	Quince Sea	
	Peach		Kakei-Wase-Hokuhu	
			Sunago-Wase	
			New Okubo	
			Schinizu-Hakuto	
			Alberta	

Source : *Agricultural Newsletter,* Ministry of Agriculture, Royal Government of Bhutan, No. 40, May-June,1994.

(a) Extension Impact and its Constraints,
(b) Farm Labour Constraints,
(c) Farm Mechanisation, and
(d) Crop Damage by Wild Life.

The approach paper prepared on a detailed investigation by a Task Force appointed for this purpose will form the guideline for future course of action.

REFERENCES

1. Bhutan, *Development in a Himalayan Kingdom 1984,* The World Bank, Washington, p. 39.
2. *LUP Newsletter,* October 1993, Ministry of Agriculture, Royal Govt. of Bhutan, Vol. 1, No. 3, p. 4.
3. *LUP Newsletter,* January 1995, Ministry of Agriculture, Vol. 3, No. 1.
4. *LUP Newsletter,* January 1994, Ministry of Agriculture, Vol. 2, No. 1, p. 10.
5. *Ibid.* p. 12.
6. A World Bank Country Study, Bhutan, Development in a Himalayan Kingdom, 1984.

7. *Statistical Year Book of Bhutan 1990*, p. 39.
8. *Ibid.*
9. *Agriculture Newsletter*, September-October 1994, Research, Extension and Irrigation Division, Ministry of Agriculture, Royal Govt. of Bhutan, No. 42, p. 1.
10. *Ibid.*, p. 11.
11. *Agriculture Newsletter*, Janunry-February 1995, No. 44, p. 4.
12. *Agriculture Newsletter*, November-December 1993, No. 37.
13. *Agriculture Newsletter*, January-February 1995, p. 4.
14. *Agriculture Newsletter*, November-December 1993, p. 7.
15. *Agriculture Newsletter*, May-June 1993, No. 34, p. 1.
16. Bhutan Country Economic Memorandum, May 1983, Vol. I, Main Report.
17. Fifth Five Year Plan 1981-1987, Main Document, Planning Commission, Royal Govt. of Bhutan.
18. Bhutan Country Economic Memorandum, May 1983, Vol. I, Main Report.
19. *Op. cit*,
20. *Op. cit.* p. 14.
21. Fifth Five Year Plan 1981-1987, Main Document.
22. *Op. cit.*
23. Seventh Five Year Plan 1992-1997, December 1991, Planning Commission, Vol. I, Main Document, p. 140.
24. *Ibid.*, p. 142.
25. *Ibid.*, p. 145.
26. *Agriculture Newsletter*, July-August 1994, No. 41, p. 6.
27. *Agriculture Newsletter*, March-April 1995, No. 45, p. 8.
28. *Agriculture Newsletter*, *Ibid.*, p. 9.
29. *Agriculture Newsletter*, January-February 1994, No. 38, p. 12.
30. *Agriculture Newsletter*, November-December 1993, No. 37, pp. 10-11.
31. *Agriculture Newsletter*, May-June 1994, No. 40, p. 10.
32. Dorji, P., 1994. 'The Agricultural Sector Policy : An Emphasis on Conservation and Sustainability in *Bhutan and Its Natural Resources*, Vikas Publishing House Pvt. Ltd., Delhi .

5

ANIMAL HUSBANDRY

— Bimalendu Bhattacharya

As an integral part of agriculture, animal husbandry takes a very important role in the life and economy of the rural population of Bhutan. The rural households irrespective of their location have a few heads of domestic animals supplementing food and earnings of the owner. The degree of dependence on the domestic animals differs widely but it is a sure means of livelihood for many, especially in the northern alpine regions where animal husbandry is virtually the sole form of economic activity. No doubt, it is an age-old occupation providing food and shelter and other essentials of daily life where consumption of animal products is an important element of the local diet in spite of the fact that the major drawback of the animals in these areas has long been low productivity due to ill health caused by diseases and lack of proper fodder in general.

As an invariable part of the farming household, the size of the herd consisting of widely different combination of animals like cattle, yak, buffalo, pig, goat, horse, donkey and mule, and chicken differs widely for any holder. Among the animals, cattle is by far the most common variety appearing in 85.6 per cent of the farm households followed by chicken with 80.1 per cent, which means that of the total number of households rearing animals at least 14.4 per cent do not have cattle and another 19.9 per cent do not possess chicken (Table 5.1). The third most dominant type of animal among the domestics is pig, reared by 53.7 per cent of the farm households followed by horse and goat belonging to less than 25 per cent of the herders. The last in the line are buffalo, yak and donkey and mule, none of them herded by more than 5 per cent of the households and among them buffalo is at the bottom of the ladder.

Table 5.1 : Percentage of Households having Livestock in Bhutan, 1988-89

Number of Livestock	Cattle	Yak	Buffalo	Pig	Goat	Sheep	Horse	Donkey and Mule	Chicken
0	14.4	97.8	98.1	46.3	74.6	90.3	73.4	95.0	19.9
1-2	12.8	—	0.9	39.0	11.7	3.8	22.4	4.7	22.8
3-4	17.1	0.3	0.5	11.0	8.9	1.9	3.1	0.2	20.8
5-9	36.2	0.2	0.5	3.4	4.1	2.0	0.9	—	21.4
10-14	12.7	—	—	0.3	0.3	0.6	—	—	8.0
15-19	3.8	0.3	—	—	—	—	—	—	3.3
20-29	2.3	0.2	—	—	—	0.3	—	—	2.5
30 & over	0. 8	0.8	—	—	—	0.2	—	—	1.3
Total	100.0	100.0	100.0	100.0	100.0	100.0	100.0	100.0	100.0

Source: Statistical Year Book of Bhutan, 1990.

Regarding the size of the herds, it is cattle which is comparatively larger for all the groups, from 3-4 heads upto 30 heads and over followed by chicken which is smaller than cattle for all the size groups, from 5-9 heads onwards. The largest size of the herd of buffalo and horse maintained by the farm households is 5-9 heads, for pig and goat 10-14, and for donkey and mule it is 3-4.

Cattle is by far the largest in number among the domestic animals in Bhutan, exceeding others, except poultry, by many times. Among the quadrupeds it is as much as nearly five times larger in number than pig, the second in successive order, maintaining the numerical strength all through the years, 1981 — 1989, for which statistics are available, followed by sheep in the third place at least for the last three years — 1987 to 1989 since when data for goat is available separately (Table 5.2). The number of each, however, has fluctuated through the years, none of them registering steady growth or decline. Besides, it is very difficult to ascertain the reasons for either rise or decline in cases where there had been a sharp increase or

Table 5.2 : Number of Livestock in Bhutan, 1981– 89

Livestock	1981	1982	1983	1984	1985	1986	1987	1988	1989
Cattle (a)	275,200	291,700	309,400	327,700	347,400	340,300	357,000	357,400	303,500
Yak	28,900	29,600	30,100	30,700	31,300	35,700	30,100	35,600	37,100
Mithun Cross	n.a.	n.a.	n.a.	n.a.	n.a.	n.a.	n.a.	n.a.	46,400
Buffalo	n.a.	n.a.	n.a.	n.a.	n.a.	n.a.	5,200	4,300	4,700
Sheep (b)	27,000	31,800	37,500	44,300	52,300	43,800	36,400	46,600	47,800
Goat	n.a.	n.a.	n.a.	n.a.	n.a.	n.a.	40,900	36,700	32,500
Pig	55,500	56,600	57,700	58,900	60,100	88,600	70,300	65,700	63,200
Horse	n.a.	n.a.	n.a.	n.a.	n.a.	23,600	26,000	25,700	26,100
Poultry	152,500	158,800	165,500	172,300	179,500	211,300	218,000	237,200	210,700

(a) Upto 1988, Mithun was included in cattle.
(b) For 1981 to 1986, the figures include both Sheep and Goat.
Source : Statistical Year Book of Bhutan, 1990.

decrease. For instance, the number of cattle rose from 340,300 to 357,000 between 1986 and 1987. In the case of Yak, a sharp reduction from 35,700 heads in 1986 to 30,100 heads in 1987 and a subsequent rise to 35,600 heads in the following year (1988) cannot be accounted for. Similar instances may be cited for other animals, both quadrupeds and bipeds for which reasons are not clear enough.

Rather, a districtwise (dzongkhagwise) distribution of these animals provides an insight into their geographical concentration or otherwise. In this case the highest concentration of domestic animals is found in Geylegphug district (Table 5.3) in the south with 12.8 per cent of the total number of heads consisting of all types, followed by Samchi having 11.7 per cent of the total and Tashigang with 10.8 per cent, both situated in the south. Together these three districts contain 35.3 per cent of the total number of livestock heads, and including Chhukha (5.8 %), Chirang (7.4 %), Daga (6.0%), Shemgang (2.8%), Mongar (6.9%) and Samdrupjongkhar (6.7%), the nine southern districts have 71% of the domestic animals concentrated in them.

Among the domestic animals, cattle and pig are the only ones found in all the districts. Cattle, the largest in numerical strength among all the animals, is mostly concentrated in the southern districts of Bhutan. Three of them — Geylegphug (13.7 %), Samchi (11.5 %) and Tashigang (11.3%), jointly possess 36.5 per cent of the cattle population and along with

Table 5.3 : Distribution of Domestic Animals in Bhutan, 1988-89

District or Dzongkhag	Cattle	Yak	Mithun Cross	Buffalo	Sheep	Goat	Pig	Horse (a)	Poultry
Thimphu	5,800	8,500	500	—	1,600	—	2,300	1,000	2,500
Zone I									
Chhukha	24,300	—	1,300	300	800	2,900	3,600	600	11,100
Ha	9,600	5,600	—	—	200	—	2,400	1,500	2,900
Paro	14,800	4,000	—	—	300	100	5,800	1,200	4,000
Samchi	34,900	—	—	1,200	4,200	10,900	2,600	—	36,400
Zone II									
Chirang	19,300	—	—	800	2,700	8,300	1,700	500	23,500
Daga	22,300	—	600	900	2,400	4,800	3,000	800	14,300
Punakha	9,400	1,900	1,700	—	100	200	3,900	1,300	6,200
Wangdi-phodrang	17,300	3,400	2,600	—	10,600	—	5,800	2,400	300
Zone III									
Bumthang	3,600	6,800	1,800	—	7,400	—	200	1,600	—
Geylegphug	41,700	—	300	1,500	4,200	2,000	4,300	600	43,100
Shemgang	12,600	—	—	—	—	600	2,500	900	5,300
Tongsa	9,800	—	3,000	—	4,200	—	700	500	3,700
Zone IV									
Lhuntshi	1,700	300	5,800	—	400	100	2,900	1,300	9,700
Mongar	17,100	—	11,500	—	200	100	5,500	2,100	16,600
Pemagatsel	5,000	—	2,900	—	—	100	2,300	1,200	5,000
Samdrup-jongkhar	19,900	100	5,400	—	500	2,400	5,200	2,200	16,200
Tashigang	34,400	6,500	9,000	—	8,000	—	8,500	6,400	9,900
BHUTAN	303,500	37,100	46,400	4,700	47,800	32,500	63,200	26,100	210,700

Source : Statistical Year Book of Bhutan, 1990.
(a) Includes Mule and Donkey.

Chhukha (8.0%), Daga (7.3%), Samdrupjongkhar (6.6 %), and Mongar (5.6%), the seven districts in the south maintain altogether 64 per cent of the total heads of the country's cattle. Its concentration in other districts is considerably less except in Wangdiphodrang (5.7 %) and Paro (4.9 %).

Unlike cattle, pig, the second largest in number among the quadrupeds, shows concentration very conspicuous in only four districts occurring in different geographical settings. They are Tashigang with 13.4 per cent of the pig population, Paro (9.2 %), Wangdiphodrang (9.2 %) and Mongar (8.7 %). The percentage of pigs is much smaller in other districts.

Sheep, third in numerical strength, is absent in at least two districts— Shemgang and Pemagatsel — but makes a high concentration in three districts— Tashigang (16.7 %), Bumthang (15.5 %) and Wangdiphodrang (22.2 %) or the combined strength of these three districts comes to about 55 per cent of the total sheep population of Bhutan.

Mithun closely following sheep in total strength is, on the other hand, more concentrated in the eastern districts (Zone IV) and is found to be absent in five districts, three of which lie in the far west. Mongar has the largest number of this animal (27.8 %), followed by Tashigang (19.4 %), Lhuntshi (12.5 %) and Samdrupjongkhar (11.6 %). Thus, the four districts of the east together maintain between them 71.3 % of the total herd of mithun in the country.

The distribution of yak is more area specific than many other animals the majority of which is concentrated in four districts, namely Thimphu (23.0 %),Bumthang (18.3 %), Tashigang (17.5 %) and Ha (15.1 %), possessing together 73.9 per cent of the total population of yak in Bhutan. It is absent in nine out of the eighteen districts, and all the nine districts constitute the southern part of the country. Yak, as one of the most hardy domestics, demand special mention. As a 'multi-purpose breed', yak is adapted to higher altitudes and enjoy the most enviable position in view of their ability in utilising higher altitude pastures where no other animal can survive. Certain physiological features, like a heavily built arota enable them to withstand the rigours of high altitude, the thin lips and strong well-jacketed incisors make it possible for them to eat soft as well as very rough vegetation. A secondary coat developed beneath the primary coat in winter, saves them from freezing cold (Ura, 1993).[1] The usefulness of yaks may be compared with that of cross bred cattle or Siri Cattle (Table 5.4), showing that yaks have a longer lactation period and a higher fat content in their milk very much favoured by the herders preferring butter to milk as it is easier to carry during seasonal movements. Besides, yak meat fetches higher price than the meat of other animals which is another reason for the herder favouring male yaks. For instance, the price of a kilogram of yak meat was Nu. 65 as against Nu. 30.80 for pork, Nu. 41.67 for mutton, and Nu. 15.42 for beef (Ura,1993).[2]

Table 5.4 : Milk Production Parameters for Different Species of Livestock in Bhutan.

Livestock Species	Average Lactation Length (days)	Average Daily Production of Milk (kg)	Yield per Lactation (kg)	Average Fat Content
Siri Cattle	240	0.75	180	4.6
Mithun Cross	253	1.25	316	7.9
Jersey Cross	246	2.00	492	4.5
Yak	280	0.70	196	6.7

Source : Ura Karma, *The Nomad's Gamble*, 1993.

The total number of goats is smaller than that of yaks but it is much more widespread in its geographical distribution than the latter though showing similar characteristics of being highly concentrated in a few districts. These are the districts of southern Bhutan — Samchi (33.5 %), Chirang (25.5 %), Daga (14.8 %), Chhukha (8.9 %) and Samdrupjongkhar (7.4 %) — together maintaining 90.1 per cent of the total goat population when it is virtually absent in six districts of Bhutan most of which lie in the north.

Horse appears to be a more common animal than mithun, yak or goat, found in all the districts except Samchi. Nearly one-fourth of the total is concentrated in one district — Tashigang (24.5 %) — and along with three other districts, such as Wangdiphodrang (9.2 %), Samdrupjongkhar (8.4 %) and Mongar (8.4 %), together they have 50.2 per cent of the total number of horses. In fact, the five districts of east Bhutan have more than half of the horse population.

Lastly comes poultry, the second largest among the domestic animals making a common occurrence in all the districts barring Bumthang, with a concentration widely different from one district to another. However, it is in the southern districts that the concentration is much higher and six districts of the south — Geylegphug (20.5 %), Samchi (17.3 %), Chirang (11.2 %), Mongar (7.9 %), Samdrupjongkhar (7.7 %) and Daga (6.8 %)—

together have 71.4 per cent of the poultry heads.

It is quite apparent that the farming population living in the districts of the south and east derive a greater benefit from a better combination of the domestication of agricultural crops and household animals.

The realisation of the fact that there was a tremendous, rather limitless potential in animal husbandry spurred activities at the official level in Bhutan and the objectives were clearly outlined at the start of the First Five Year Plan (1961) as follows :

(a) Prevention of livestock diseases and epidemics taking a heavy toll on livestock population;
(b) Introduction of better yield and better draft animals into the country for improving productivity and draft power of the indigenous varieties through cross breeding;
(c) Education and training of the farmers in modern techniques of animal rearing to make this livelihood more remunerative in the long run.

With this aim in view, the achievements of the First Plan (1961-66) primarily remained confined to the establishment of (i) a few Livestock Farms to introduce, multiply and acclimatise exotic animals for distribution to the farmers[3] and demonstrate to the farmers the modern methods and techniques applicable to animal rearing, and (ii) the establishment of veterinary dispensaries to prevent cattle epidemics. Two Livestock Farms and four Veterinary Dispensaries came up during this period. The Livestock Farms were established at Samchi and at Bidung in Tashigang. The farm at Samchi occupying 54 acres of land started its operation with 8 Haryana cows and 2 bulls, 8 Murrah buffaloes with 2 bulls from Calcutta to which 7 Jersey cattle were added later as a gift from Sikkim. The number of cattle increased to 80 by multiplication by the end of the Plan period, and Poultry and Piggery Sections were also introduced. The second Livestock Farm at Bidur started with 14 Mithun Cross cows, 6 Jersey Cross cows and 2 pure Jersey bulls. Besides, a Piggery Unit and a Mule Breeding Centre were added to this unit. The First Plan period further witnessed the opening up of a Sheep Breeding Farm at Mera/Khaling in 1965 with 100 Rampur Bushahir sheep imported from Himachal Pradesh in India which were multiplied into 150 heads by the end of the First Plan. Besides, a Pisciculture Unit for breeding trout was started at Ha during this period,

Plate 9 : Mithun with other cattle

Plate 10 : Thimphu Bazar

producing 12,000 fingerlings and later on released them into the seven rivers of the country.

During the Plan period, four veterinary dispensaries were established at Samchi, Thimphu, Paro and Sarbhang which, apart from treatment, carried out vaccination for 32,675 heads of livestock.[4]

During the Second Five Year Plan period (1966—71), three more Livestock Farms came up. The first of them was at Wangchutaba in Thimphu for cattle, pigs and poultry; the second one at Lingmeythang for cattle, pigs and poultry, and the third at Thromang for mithun breeding. At the same time five new Veterinary Dispensaries were set up respectively at Phuntsholing, Wangdiphodrang , Deothang, Tashigang and Mongar. In all, 64,743 heads of animals were treated and 303,500 heads were vaccinated during the Second Plan period. Further, in order to check the death of yak calves, a Yak Disease Eradication Scheme was launched. In terms of notable achievements, 132 bull/bull calves were supplied to the farmers for cross breeding; 1,145 heads of pig/piglets were distributed among the farmers for quality upgradation; the farmers were supplied with 48 mithuns, 2,477 poultry birds and 90 sheep for cross breeding; and 34,000 trout fingerlings were released into the rivers and lakes.

The major objectives of the Third Plan (1971—76) were strengthening and equipping the farms and the dispensaries established during the earlier plan periods for deriving maximum benefit. A Disease Investigation Laboratory set up at this time at Thimphu proved very useful in diagnosing and controlling parasitic diseases among the animals. The opening up of a Veterinary Training Institute at Wangchutaba in Thimphu and two more Livestock Farms respectively at Ha and Sarbhang, 13 Veterinary Dispensaries/Sub-Dispensaries repectively at Ha, Punakha, Tongsa, Dorokha, Sibsu, Lhuntshi, Daifam, Shemgang, Dokar, Damphu, Geylegphug and Shumar along with the opening up of Cattle Bull Centres attached to the Veterinary Dispensaries at Thimphu, Paro, Wangdiphodrang, Mongar and Phuntsholing for extending cross breeding facilities to the farmers came to be some of the major achievements in the Third Plan. Another important step taken during this period was the establishment of a National Sheep and Yak Development Project at Bumthang for upgrading the productivity on the one hand and arresting the rate of mortality on the other. Another Pisciculture Unit was established at Wangchutaba for trout breeding and 233,448 fingerlings were released during the plan period. The Mithun Breeding Farm operating at Thromang in eastern Bhutan was shifted to Chhukha, providing a more suitable climatic and topographic environment for Mithun culture. During this period, 396,171 animals were vaccinated. Rinderpest, a deadly disease among the cattle, was finally controlled in 1971 and 150,968 animals were treated at the Veterinary Dispensaries.[5]

From the Third Plan, the improvement of the indigenous stock through cross breeding with imported exotic stock became the primary objective of the animal husbandry programme which was carried out in the government farms. This, covering all types of animals, was taken up as Intensive Livestock Development Project, setting up 12 Livestock Breeding Farms and 19 Bull Centres in the central and southern parts of the country. This resulted in a substantial increase in cross-bred population of cattle, pigs, sheep and poultry and a greater popularity of them with the farmers who by then were more alive to scientific techniques. For the first time a dairy development programme in conjunction with milk collection centres was initiated in the urban areas. A Grass Seed Production Centre was set up to help pasture development in the northern alpine region. The veterinary service developed till this time included a hospital and a diagnostic laboratory at Thimphu, 28 dispensaries and 19 sub-dispensaries in the major livestock rearing areas. Among the notable achievements had been the control of tapeworm cyst infestation in yaks, leading to a remarkable fall in the mortality rate of them, forming the primary basis of livelihood for the inhabitants of the alpine regions.[6]

Thus, the major emphasis throughout these twenty years of the four Five Year Plans has been on the 'induction of hybrid stocks and their sale to farmers at subsidised rates in order to improve the productivity of local animals; and the establishment of an extension and veterinary infrastructure to control livestock diseases and train farmers in modern practice'.[7]

In summary, one of the ultimate objectives behind all these programmes was to improve the intake of nutritional food for the Bhutanese people comprising high amount of protein 'half of which should be of animal origin'.[8] For instance, when per capita consumption of milk in developed countries was 400 kg a year, in Bhutan it was 30 kg even in the late seventies.[9]

The results obtained from four Five Year Plans made a sound footing for further activities. For instance, in the Fifth Plan, the major focus was on the following aspects with a greater allotment of funds for this purpose:

(a) Improvement in the productivity of livestock through an integrated programme on scientific breeding, better feeding, improved management and an effective animal health programme;

(b) To supplement the income of the livestock farmers through a better marketing system;

(c) To increase the production of livestock products for improving the nutritional intake of the people and attain self-sufficiency in livestock requirement.

The major thrust in this programme was on improving the genetic quality of the cattle through cross-breeding of the local breeds with the exotic breeds since 40 per cent of the former was unproductive. This did not, however, mean complete removal of the local breeds, particularly the good ones like the Siri breed, the best of which were to be used for cross-breeding in a selective manner. Along with this, the cross-breeding of local Siri cows with mithun bulls was to continue and to ensure a steady supply of the latter, the existing mithun breeding farms at Arong and Chhukha were going to be strengthened. Further, the Jersey Cattle Farm at Samchi was to be enlarged to have 80 cows and to produce 20 pure bred Jersey bulls every year. Two bull-calf rearing farms were to be established at Thungti and Tala. Over and above this, the cross-breeding programme with Swiss bulls already taking place in Bumthang district was to be extended further through a network of 40 natural breeding centres. In order to improve the quality of the local buffaloes, it was proposed to distribute 17 Surti buffaloes to the southern districts.[10]

In Bhutan, milk traditionally converted into butter and cheese because of the adverse geographical and transportational conditions, made fluid milk almost a scarce material but with urban development in recent decades its demand is on the increase for which collecting and marketing units started functioning at places like Phuntsholing, Samchi, Geylegphug, Gesu and Thimphu. The production of milk at the beginning of the Fifth Plan (1981—87) was 16.50 million litres which was to be increased to 26.50 million litres by the end of the Plan period and for achieving this, greater efforts were to be made in organising collection and marketing of milk on a greater scale at Thimphu, Geylegphug, Gedu and Samdrupjongkhar.[11]

The growing demands for poultry in the rural and urban areas were to be met by commercial production of eggs and broilers for which a unit was to be developed near Thimphu with a production capacity of 3,000

Plate 11 : Thimphu Bazar

Plate 12 : Thimphu Bazar

table eggs daily and 1,000 broilers every month and a project with a production target of 40,000 broilers per year. The poultry farms at Paro, Sarbhang and Lingmethang were to be strengthened for breeding, rearing and the supply of cross-bred birds to the rural areas generating an extra source of income.

The introduction of genetically superior breeds of pigs marketable at an early age being the main objective, cross-breeding of the indigenous pigs with the exotic ones was to continue and for attaining self-sufficiency in quality boars, the pig farms at Paro, Sarbhang and Lingmethang were to be expanded during the plan period. At the same time, the Central Pig Breeding-cum-Pork Processing Plant at Wangchutaba near Thimphu was to go through further expansion.

Efforts of the Government was to be directed during the plan period towards increasing and improving the number and the quality of sheep in the rural areas. For this purpose, cross-breeding will continue to form a single pool of Merino-type at Bumthang and Tashiyangphug farms which was given top priority and the farms were going to be strengthened further for the production of pure-bred rams for distribution in the neighbouring areas. A large sheep breeding farm was to be established in western Bhutan to upgrade the local sheep population in the south and the western parts of the country.

Improvement in the quality and increase in the number of yaks entailed a programme for encouraging the exchange of yaks from one rearing region to another and thus avoid in-breeding while the yak breeding farms at Shingkhar and Shabi Tongerfog were to be used for selective breeding among yaks and for providing cross-breeding facilities with Jersey breeds.

Ponies and mules have a great traditional role in transport in the mountainous country of Bhutan and to check the degenerating quality of the local ponies, otherwise extremely suitable to local conditions, a stud farm of Halflinger mares was to be developed at Bumthang for cross-breeding with the local types which could help 40-50 stud centres in other areas. And for improving the quality of mule, Italian stallion were to be imported during the plan period.

Animal Feed and Animal Health Care

Feeding animals has turned out to be a growing problem with increasing human encroachment into the feeding grounds. Any growth in the animal population makes a greater demand on the available food provided mainly by the pasture land amounting to about 1,564 sq km, or 3.9 per cent of the total land area of Bhutan[12] of which about 40 per cent is considered to be 'unproductive and low yielding'. Both private and community pasture land were to be developed for which the Forest Department was to take initiative during the plan period providing rhizomes, legumes, fodder plants and technical advice to the farmers. The pasture and fodder tree plantation was to continue as a part of social forestry programme and the Grass Seed Multiplication Centre at Sarbithang was to be strengthened to produce more fodder seeds. Along with this, the systems of developing perennial pasture was to be encouraged and demonstrated for rotational grazing. Steps were going to be taken for the manufacture of animal feeds at local levels and establish a feed mixing plant and godown at Phuntsholing making use of locally available feed like maize, rice bran, rice polish and industrial by-products.[13]

Implementation of preventive treatment was given top priority in the Fifth Plan with a programme for the expansion of the Veterinary Diagnostic Laboratory at Sarbithang along with the addition of a new feed analysis section at this laboratory. A Veterinary Vaccine Production Unit was also proposed to be established producing initially vaccines against HS, B.Q. and Anthrax infections and acting as a storage depot for all other biological requirements it will eventually develop into a fully-fledged 'biological products' station. The Veterinary Hospital at Thimphu was to be strengthened further when all the seventeen Veterinary Dispensaries in the districts were to be upgraded into hospitals.

Objectives and Development during the Sixth and Seventh Plan Period

The objectives of the successive two plans — the Sixth Plan (1987—1992) and the Seventh Plan (1992—1997) — were more or less similar to those of the earlier plans. During the Sixth Plan period, new breeds have been produced at the Department of Animal Husbandry Farms and were distributed among the farmers. The adoption of the improved breeds obtained through cross-breeding had been higher in the ADP (Area Development Programme) areas where the extension programmes consisting of high level technical support have been more intensive. The pasture and improved grazing management programmes had also been implemented and a Fodder Research Master

Plan had been initiated to coordinate the activities of the fodder research stations. A particularly significant feature is that privatisation is gaining ground in the activities of the animal husbandry sector. This is exemplified by the privatisation of the poultry section of the Central Pig and Poultry Breeding Centre at Wangchutba. The milk distribution and marketing system in Thimphu had also been privatised during this period while the privatisation of the Regional Mithun Farm in Chhukha and the Milk Processing Plant in Phuntsholing was under consideration. During this period, marketing cooperative societies came up, increasing opportunities for the marketing of milk in the rural areas of Bhutan. There were only a few slaughter houses in Bhutan, located close to the border with India and the livestock from the interior is exported to India. There are four quarantine stations along the border at Samchi, Phuntsholing, Geylegphug and Samdrupjongkhar which are responsible for quarantining animals for 15 days before entry into the country.

In the Seventh Plan the major thrust is on "increasing productivity through breed improvement and processing of livestock products to increase incomes rather than on increasing livestock numbers".[14] The growing problems of feeding animals is likely to worsen the situation further if the number of mouths to be fed increase steadily which eventually will have a deterrent effect on quality. Hence, to maintain and improve the quality of livestock, breed improvement came to be a major component of livestock development programme for which artificial insemination facilities will be strengthened and new centres will be established in specific areas for this purpose. Mobile insemination services will be established where feasible and distribution of cross-bred stock from the network of breeding farms will continue. Simultaneously, artificial insemination facilities will be introduced for other livestock. Milk processing and marketing centres will be established for encouraging milk production and extension services will be provided to the rural areas through the formation of cooperative societies. The sheep production at the breeding centres of Bumthang, Tongsa, Wangdiphodrang and Tashigang will be expanded further for producing improved variety of sheep for wool and this will form the basis for wool processing and weaving industry. The Government Animal Husbandry Farm will supply improved breeds to the farmers and this activity will be expanded further to meet the growing requirements in the countryside.

The commercial development of dairy, piggery and fish farms will be made for providing urban areas with livestock products. The identification of suitable species for agro-forestry and fodder production done under pasture development programme will continue for improving animal nutrition and inter-cropping of leguminous fodder providing additional feed for animals and enhancing soil fertility will be encouraged. The fodder programme will further incorporate the production of fodder tree for distribution under social forestry programme. Over and above this, the existing facilities for animal health services will be strengthened further and expanded. An Epidemiological Unit will be set up at Sarbithang to study the pattern of the outbreak of epidemics in the country. At present, matitis is one of the major health problems in dairy cattle and the high incidence of the disease in Bhutan is a cause for great concern for which poor sanitation and unhygienic milking procedures are cited to be the main reasons.[15] Foot and mouth disease ranking as the number one livestock disease in Bhutan,[16] however, is thought to be more difficult to eradicate through vaccination alone unless the farmers become more alert. For combating these diseases on a greater scale, the Seventh Plan makes provision for further strengthening of the Zonal Laboratories by providing additional facilities for immediate detection and diagnosis of diseases in the rural areas. The production as well as distribution of vaccines will be upgraded with increase in production, giving greater attention to 'surveying, control procedures, and quarantine mechanism to prevent the spread of animal diseases'.[17] The voluntary village animal health workers in the rural areas will be trained through Mobile Services to be introduced during the plan period.

According to the global data bank at F.A.O., Rome, 36 per cent of the animal breeds (excluding poultry) of the world are in Asia and Pacific regions, making them the most richly endowed area in terms of domestic animal diversity but some of them are already at the risk of becoming extinct. The exigency of the situation has resulted in the emergence of the animal genetic resources project known as 'The Conservation and Use of Animal Genetic Resources in Asia and the Pacific', which will cover twelve countries and Bhutan is one of them. The project will focus on obtaining 'better population data' on animals and get involved in the 'development of schemes to improve the efficiency of indigenous breeds within the environment in their own habitat and preservation either as live animals or in cryogenic form, or preferably both. The project is expected to ensure that "bio-diversity, at least for animals, is fully utilised and maintained in order to provide food, fibre and fuel for future generations of mankind."[18]

The proposals likely to emanate from a study of this magnitude are expected to contribute significantly to the conservation of animal genetic resources and the endangered species from imminent extinction.

REFERENCES

1. Ura, K., 1993, *The Nomad's Gamble*, National Environment Commission, Thimphu, Bhutan.
2. *Ibid.*, p.15.
3. Projection, Animal Husbandry Department (1961-1981), Royal Govt. of Bhutan.
4. *Op. cit.*
5. *Op. cit.*
6. Bhutan, Country Economic Memorandum, May 1983, Vol. I, Main Report.
7. *Op. cit.*
8. Review on Animal Husbandry Activities in Bhutan, Animal Husbandry Department, Royal Govt. of Bhutan.
9. *Op. cit.*
10. Bhutan, Country Economic Memorandum, May 1983, Vol. I, Main Report, pp. 108-119.
11. *Ibid.*, p. 110.
12. *LUP Newsletter*, January 1995, Ministry of Agriculture, Royal Govt. of Bhutan, Vol. 3, No. 1.
13. Bhutan, Country Economic Memorandum, May 1983, Vol. I, Main Report, p. 114.
14. Seventh Five Year Plan 1992-1997, December 1991, Planning Commission, p. 115.
15. *Agriculture Newsletter*, September-October 1994, No. 42.
16. *Agriculture Newsletter*, April-May 1995, No. 45.
17. Seventh Five Year Plan 1992-1997, December 1991, Planning Commission.
18. *Agriculture Newsletter*, November-December 1994, No. 43.

6

FORESTRY

— Bimalendu Bhattacharya

FORESTS spread over almost 72.5 per cent (76.4 % including pastures) or 29,045 sq km (30,609 sq km including pastures) of the total land area constitute the most important renewable resource of Bhutan, playing a vital role in ecological balance and economic development of the country. It has partly been the basis of livelihood for the Bhutanese people from time immemorial and forms today the very foundation of the relative prosperity of the country. According to *Druk Forestry News,* about 68.66 per cent or 26,963 sq km land area of the country is under forests, but this computed by P.N.Gupta based on remote sensing in 1992 is not considered to be authentic as "he (Gupta) did a very quick job within a deadline whereby he could not have full coverage of the country and he did not have time for field verification".[1] Instead, the figures given by *LUP News* are taken to be more accurate and uptodate. As mentioned earlier, the wide variation in physiographic and climatic factors is distinctly reflected in the character of the natural vegetation and forests ranging between alpine and sub-tropical types occur with a great variety of plant species. The great diversity of the plant kingdom finds support in the following statement :

"Bhutan's diverse fauna and flora include an estimated 770 bird species and at least 5,000 species of vascular plants. In addition, more than 60 per cent of the species endemic to the Eastern Himalayas are found in Bhutan".[2]

Figures presented in the Table 6.1 show that among the forests the broad variety dominates in areal occupance, constituting about 45.1 per cent (including land under broad-leaf plantation) of the total forest land. Next to it in importance is the coniferous group with an areal occupance of 26.5 per cent and including new plantations it comes to 26.6 per cent of the country's land area which in terms of the total forest land is 34.8 per cent while these two groups along with the mixed forests of the 'broad-leaf and conifer' hold together 64.4 per cent of Bhutan's total land area, comprising about 84.3 per cent of Bhutan's forest land.

Table 6.1 : Land under Forests in Bhutan

(area in km and % of total area)

Category or Sub-Class or Class of Forests	Area (km^2)	As % of the Total Area	As % of the Total Forest Area
Fir	3,453	8.6	11.3
Mixed Conifer	4,868	12.1	15.9
Blue Pine	1,286	3.2	4.2
Chir Pine	1,009	2.5	3.3
Conifer Forest	10,616	26.5	34.7
Broadleaf and Conifer	1,358	3.4	4.4
Broadleaf Forest	13,749	34.3	44.9
Conifer Plantation	20	0.1	0.1
Broadleaf Plantation	44	0.1	0.2
Forest Plantation	64	0.1	
Scrub Forest	3,258	8.1	10.6
FOREST	29,045	72.5	94.9
Natural Pastures	1,553	3.9	5.1
Improved Pastures	11	0.0	00.0
TOTAL	30,609	100.0	100.0

Source : LUP News, Vol. 3, No. 1, January 1995.

In general, the broad-leaved tropical hard woods dominate up to an altitude of about 1,000 m, beyond which the temperate forests of pine and oak reign. In the inner valleys oak and other deciduous trees gradually give way to conifers with increasing altitude. Pine and oaks are mainly concentrated on the southern slopes with greater rainfall 3,000 m and upto about 4,000 m which is the tree line, very valuable timber-yielding species of pine, spruce and fir occur. Above the tree line the vegetation grades into scrub and grass, representing the grazing land of animals which covering altogether 12 per cent of the total land of the country constitute almost 15.7 per cent of the forest land to play an important role in the national economy.

Among the temperate forests, *Fir* forms the highest one in altitude in Bhutan, growing from 3,000 m up to the tree line at 4,000 m. At high altitudes it makes pure stands or mixed with junipers, often with rhododendrons as undergrowths. It prefers moist areas and is found, mixed with other conifer varieties at altitudes 3,000—3,500 m. It is usually 30-40 m tall below 3,600 m but becomes shorter higher up, grazed by yak in winter when no grass is available, making it a problem for new plantations. The wood is soft and light, used for particle board and for roof shingles.[3] *Blue Pine,* on the other hand, is a tree of comparatively lower elevations, growing mainly on the lower valley slopes of western and central Bhutan at altitudes of 1,700—3,300 m where it occurs as a pure stand while higher up it appears with other conifers and oaks. It is replaced by a typical Bhutanese pine (*pinus Bhutanica*) in the east which is very similar to blue pine. It regenerates easily, invading pastures and orchards. Blue Pine may grow up to 50—55 m in height in favourable conditions and is the most popular tree in Bhutan for construction purpose, used for building houses, dzongs, and the construction of furniture while the leaves find use as litter in the cowshed and as manure. Blue pine timber has export value. *Cypress* considered as a sacred tree is locally called 'Tsenden', the national tree of Bhutan, grows between 2,550 and 3,000 m but is also planted on lower elevations and thrive well on steep limestone areas, and as a sacred tree are found to be common around religious buildings. It is tall, an impressive tree grows up to 50 m and the timber is good for construction works, and particularly used for building temples while the wood and branches are burned as incense at the entrance of monasteries and temples. *Spruce,* a very tall tree of 50—55 m , is another very common type of conifer growing well at altitudes 2,400—3,600 m, occurring in pure stands and found sometimes mixed with other conifers and oaks. It is the second most preferred tree after blue pine for its timber growing in the temperate forests going extensively into the construction of buildings and ordinary houses and as shingles for roofing. *Larch,* the only conifer to shed its leaves in winter, grows at elevations between 3,000 and 3,800 m mixed with other conifers like spruce, hemlock and fir. The wood has no specific use. *Hemlock*, a coniferous tree of large size, growing up to 40 m, is found at altitudes of 2,400—3,300 m from the western to the eastern parts of Bhutan both as pure stand or mixed with other conifers like blue pine, spruce or fir, or occur in the mixed conifer zone with oaks having rhododendrons and bamboo as the undergrowth. The tree is felled for commercial logging and the timber is also used for constructions. Similarly, *Juniper*, another conifer is also a high altitude variety growing between 3,200 and 4,700 m in forests in the western as well as the eastern parts. Growing up to about 15 m in height, its wood is popularly used for funeral pyres and the twigs and leaves are used as incense. The young juniper is not usually grazed by yak and cattle.

Among the broadleaved species of the temperate forests there are two types of *birch* in Bhutan of which Betula Alnoids grows at medium altitudes, and Betula Utilis grows at higher altitudes of 3,100-4,200 m mixed with firs, hemlock and rhododendrons. It is of shorter height, reaching up to 15 m. The wood is used on a small scale as timber for house construction and manufacture of agricultural implements while the water-proof bark is sometimes used as paper and in treating wounds. *Maple,* another tall tree of coniferous group, consisting of a few varieties in Bhutan, appears in moist forests between 2,100 and 3,600 m mixed with other conifer and broadleaved species. Maple wood find use in making farm equipments, broom sticks and furniture.

Magnolia growing at altitudes 2,000—3,100 m in moist area forests, is a broadleaved tree getting bare of leaves in winter. The wood being weak, has no commercial use. Of the two types of *Oaks*, one (*quercus semecarpifolia*) is very common all over Bhutan, found between 2,400 and 3,600 m, becoming a tall tree, 30 m high, under moist conditions. An evergreen oak, it mostly occurs with conifers or other broadleaved species. The wood is preferred as firewood while the green leaves make fodder for animals throughout the year. The other type (*quercus griffithii*) comprising the shorter variety, 25 m tall, grows in temperate and warm temperate areas at 1,300—2,500 m, sheds its leaves before winter which is used as litter in cowsheds and mixed with cowdung make good manure. The hard wood is used as fuel. *Poplar*, a fast growing plant as tall as 25 m, occurs in the temperate area of western and central Bhutan between 2,100 and 3,000 m. The wood is used for making match-sticks and packing boxes. *Walnut*, growing at 1,500 to 2,500 m in mixed broadleaved forests, is found in western and central Bhutan, being more common in eastern Bhutan. Nearly 25 m tall, economically it is a very valuable tree for its fruit, and the wood is highly valued for its durability, mostly used for making furniture and a black dye is made out of its bark. *Rhododendrons* occur in between 1,500 and 3,300 m and either as pure stands or mixed with other species above 2,300 m. There are 50 different types of rhododendrons in Bhutan which grow up to 10 m in height. Its wood is used for the manufacture of particle board, traditional bowls and knife handles. The wood is also used as firewood. The flower has some medicinal value for curing dysentery and young leaves are put on the forehead to relieve one from headache.

These are some of the most widely found trees in the temperate and sub-alpine vegetation zones of Bhutan. There is, however, one particular vegetation common to almost all climatic zones and of great economic value which is *bamboo*, providing constructional, weaving and fencing materials, animal fodder and it is one of the most profitable plants that marginal land can sustain, offering greater protection against soil erosion.

'Lho menjong', another name for Bhutan, means the 'land of medicine of the south', in which 'south', according to some authors 'connotes land or country lying south of Tibet, in fact, called the Southern Country'. The great diversity of plants is a repository of a large number of medicinal plants though the inventory of them is yet far from complete. Among the most important species occurring according to geographical locations are as follows (part of them used as medicine mentioned within brackets) :

(a) *Tropical-Subtropical Zone*[4]

1. Terminalia bellerica (fruit)
2. Terminalia Chebula (fruit)
3. Aquillaria Aquallocha (heart wood)
4. Holarrhona Antidysenterica (bark)
5. Rauvolfia Serpentina (root)
6. Pipper Oblongum (fruit)

(b) Temperate Zone

7. Berberis Aristala (bask)
8. Datura Stramonium (seed extract)
9. Dichroa Febrifuga (root)
10. Panax Pseudo-Ginseng (tuber)

11. Rosa Microphila (flower)

(c) Sub-Alpine Zone

12. Thalictrum Poliolosum (root)
13. Phodophyllum Emodi (root/fruit)
14. Aster Neo-Elegans (whole plant)
15. Nardosh Tachys Jatanusi (root)
16. Rheum Acuminata (root).

(d) Alpine Zone

17. Aconitum Nepalensis (tuber)
18. Delphinium Brunonianum (whole plant)
19. Picrorhiza Kurrooa (root)
20. Rheum Nobite (root)
21. Meconopsis Horridula (whole plant)
22. Codonopsis Clematidea (whole plant)
23. Fretelaria Roylie (bulb)
24. Codyceps Sinensis (whole plant).

It is reported that all these plants are dwindling in area and threatened with extinction primarily in the south because of the indiscriminate felling between 1955 and 1975 for government revenue. Besides, in spite of strict government regulations implemented in the later years, illegal collection or poaching along the southern and northern borders has caused great harm to the existence of particular plants. Traditional collection of plants for use as incense etc. by or for local doctors (Dungtsho) in remote areas may not be large in amount but it is difficult to ascertain the quantum of actual damage done to the ecological settings by such activities. Similarly, the collection of medicinal plants made annually in every season for indigenous hospitals supervised by trained 'Dungtshos' (local doctors) cannot be considered to be a very safe process as local hands are employed for this purpose.

Forest Policy and Achievements during the Period of Five Year Plans

Forests having been the largest single contributor to the national exchequer in the past, still hold the greatest potential for its economic development under proper, scientific management but this is largely lacking as evident in the substantial denudation of hill slopes and destruction of forest wealth resulting from unscientific working in the recent years. The first priority of the early plans had been 'undertaking a pre-investment survey of the forest resources and to establish an administrative infrastructure' to provide a basis for a scientific management. In sequence, a National Forest Policy was adopted in 1974 to formulate principles for conservation, development, training and administration as a consequence of which all commercial felling by private contractors was banned in 1979 and the export of timber was nationalised.[5]

The major thrust during the twenty years of the first four Five Year Plans (1961—81) was on establishing a physical, administrative and technical infrastructure, with trained personnel to undertake the future programme of forest management. In pursuance of this a pre-investment survey of forest resources started in 1974 which covered 3,000 sq km of the forest area and provided an estimate of about 480 million cubic metres of growing stock volume including coniferous and non-coniferous with a net annual yield of wood approximating 13.3 million cubic metres. The survey continued during the third and fourth plan periods made a coverage of 8,140 sq km for obtaining primary data on forest resources besides spelling out specific rules for felling and making detailed management plans including silvicultural requirements, but scientific harvesting did not progress appreciably due to manpower shortages and other constraints. Not more than 0.2 million cubic metres of timber was harvested in 1980-81 and to improve the situation, greater emphasis was laid on mechanised logging. Three such logging centres were opened by the end of the Fourth Plan and about 100 km of forest roads had been completed by this time providing greater access.

The preservation of the forest cover in a mountainous country like Bhutan is essential to conserve the environment and ecological balance. The management of forests received greater attention during this period and a programme of afforestation was started to restore the heavy damage already done by indiscriminate felling in accessible areas, encroachment made by isolated human habitations, shifting cultivation, and natural causes like forest fire, soil erosion etc. About 18,600 acres of land was brought under new plantations by the end of 1980. At this time a programme on social forestry was initiated to meet the local requirements of firewood and timber so that further interference with the forest areas will cease to be a serious threat to the national wealth. The demand for firewood, timber for house-building, fencing of fields and grazing makes a heavy toll on the forests. The size of an average dwelling house, particularly in coniferous areas, is rather large, requiring a large amount of timber, which is a recurrent demand because it is needed to be replaced every 60 years while the wooden roofing barely lasts 6 years (Ura, 1988)[6], leading to inestimable loss of this valuable natural resource. As a part of the conservation process, nine wild life sanctuaries came up to conserve rare species on the way to extinction.

Since commercial exploitation of the forests, as long as it is done on a scientific manner, is not in conflict with a scientific management and utilisation of forest resources, the major objectives of the forest policy in the 1980s were as follows :

(a) Conservation and augmentation of forest resources;
(b) Scientific management and exploitation of forests for meeting people's requirements for forest products, promotion of forest based industries, particularly those with a high value added content, and increase in the export of forest products;
(c) Preservation of the ecological system with emphasis on wild life.

The mapping and marking out the boundaries of forest areas had been completed for about 11,000 sq m by the end of the fourth plan and an additional 10,000 sq m was to be covered during the Fifth Plan period. Natural regeneration was to be promoted as a part of afforestation programme supplemented by new plantations to cover 4,324 acres during this period. In order to encourage social forestry, each farmer was to plant 10 plants on his own land obtaining seedlings free from the government for which over 400,000 seedlings were to be distributed. Silviculture acquired special significance involving the maintenance of growing stock primarily through timber cutting and trimming of natural forests, and old plantations. Forest fires, an occasional phenomenon, making serious damages were going to be tackled with greater efficiency and during the Sixth Plan period the incidence of forest fires has been reduced to a great extent. The existing wild life sanctuaries were going to be equipped with improved infrastructural facilities. Two forest based industries were going to be established during the plan period. One of them was the Integrated Mechanical Woodbased Industrial Complex at Gedu, involving a veneer tory in the first phase and a plywood manufacturing plant in the second phase with an eye to export the products to India. The second unit was a Graded Particle Board Factory also located at Gedu producing laminated, veneered and other particle boards.

The policy of the Royal Government of Bhutan has persistently been putting higher priority to the protection of forests than to commercial exploitation which has brought into force laws and regulations restricting grazing in critical watersheds and protected forests, the banning of logging on steeper slopes, and restriction of conversion of forest lands into other uses. During the Sixth Plan period 'The Bhutan Forestry Development Project, Phase I and Phase II' integrated felling, extraction, plantation and nursery development, staff training and bark beetle control under one umbrella in Bumthang. This project contributed in a large measure to the development of departmental capacity in planning, logging, reforestation and disease control in both hardwood and softwood areas. The 'Integrated Forest Development Project', initiated in Bumthang district, included forestry units and a silviculture training centre.[7] Tree planting having been encouraged since 1979 by a royal decree, had been taken up as a large afforestation programme, covering 18,000 hectares by the end of the Sixth Five Year Plan and inducing rural people to plant trees for providing fuelwood, fodder, construction materials for their own use besides helping in conservation.

However, it was felt that throughout the understanding of the farmer's needs and local forest uses has not been sufficient. Hence, the Department of Forests recognised the importance of involving rural communities in forest activities which became the major strategy for the Seventh Plan. The collection and analysis of data was to continue and was going to be integrated into a Geographical Information System to provide the basis for forest management and land use planning. The afforestation programme was to continue with a fresh target of 2,500 hectares per year in order to improve soil and water conservation and, at the same time, to meet the rural requirements. The community participation in social forestry programme would be encouraged for which hence forward the Afforestation Division would take charge of the Seed Storage Centre responsible for the distribution of seeds for the afforestation programme. According to the MPFD (Master Plan for Forestry Development), of the 230,000 hectares identified as degraded forests, only 32 per cent required afforestation while 68 per cent may be converted to agricultural land including agroforestry. A greater emphasis was

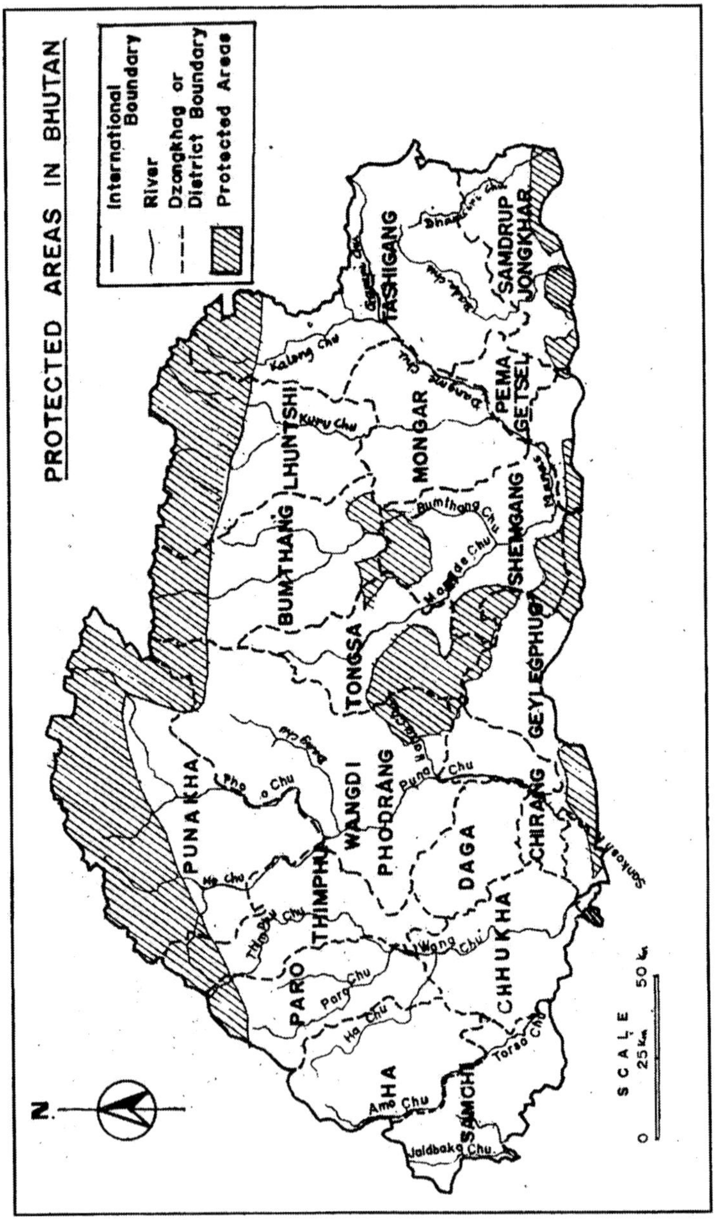

Fig. 6.1 : Protected Areas in Bhutan

emphasis was laid on the development of management plans, making sustainable utilisation of forests by communities and commercial enterprises more feasible. In the forest protection programme, the activities for fire prevention were strengthened and extended further through greater publicity and governmental help. A Protection Cell has been established to offer help to farmers in controlling pests and diseases. The development of management plans for protected areas, including Jigme Dorji stretching from west to east covering the northernmost part of Bhutan (Fig 6.1) were to continue during the Seventh Plan period though it is maintained that the boundaries of these reserves were going to be revised as, at present, they include a high proportion of alpine and tropical habitats but lack temperate and sub-alpine habitats.[8]

One of the major issues getting priority for consideration in recent years is Indigenous Forest Management which incorporates the idea of collecting information and acquiring a better understanding about the role of local population in the management of forest resources in the past and the present, and the correlation between forests and the local livelihood system.[9] Another major issue is forest fire emerging as one of the biggest threats to the forest resource. Blue Pine, Chir Pine, Mixed Conifer, Broadleaf with Conifer, plantations and degraded forests comprising about 40 per cent of the forest area are extremely prone to fires causing irreparable damage. The incidence of these fires is generally high in winter months and as investigated, hundred per cent of them are caused by human action, directly or indirectly, mentioned as debris (from the field) burning responsible for 40 per cent of the forest fires every year, burning for the growth of new grass as feed for cattle making 30 per cent, camp fire, cooking fire, room heating and road maintenance work fire causing 25 per cent, and smokers igniting 5 per cent of the forest fires occurring every year (Chhetri, 1992).[10] The eastern zone, receiving very little rainfall, happens to be the high fire risk zone. Similarly, the western region is also high fire risk zone due to low rainfall. In contrast, comparatively higher amount of precipitation and a slightly moist forest floor make the central and the southern region less susceptible to fire.

An idea may be obtained from the following table about the production of logwood over the years, forming a major item of export (Table 6.2). The data for a period of seven years (1982-83 to 1988-89) shows that commercial logging comprising the larger volume of total logging, has gained greater importance since 1983-84, attaining except in the financial year of 1985-86 an overwhelming importance in

1987-88 and 1988-89, indicating the fact that private logging is no longer encouraged.

Table 6.2 : Production of Logwood in Bhutan

('000 cubic metres)

	1982-83	1983-84	1984-85	1985-86	1986-87	1987-88 (a)	1988-89
Commercial Logging Bhutan Logging Corporation	34.0 (37.8)	59.7 (65.2)	100.8 (65.2)	59.1 (43.1)	118.1 (79.2)	76.9 (91.7)	83.4 (93.3)
Logging for House Construction and Public Works	55.8 (62.2)	31.8 (34.8)	54.3 (34.8)	77.7 (56.9)	31.0 (20.8)	6.8 (8.3)	6.0 (6.7)
Total	89.8	91.5	155.1	136.8	149.1	83.7	89.4

Source : *Statistical Year Book of Bhutan,* 1990.
(a) 1987-88 refers to 15 months, April to June.
Note : Percentage shown within brackets.

REFERENCES

1. *Druk Forestry News,* March 1993, Department of Forestry, Ministry of Agriculture, Royal Govt. of Bhutan, Vol. 1, No. 1.
2. United Nations Development Programme, 1993, Trust Fund for Environmental Conservation in Bhutan, Royal Govt. of Bhutan, WWF.
3. Common Trees in the Temperate Forest of Bhutan, 1992, National Environment Commission, Thimphu, Bhutan.
4. Nowrang, R., March 1989, 'Medicinal Plants and their present status in Bhutan' in *Tsenden : A General Publication on Forestry in Bhutan,* Vol. 1, No. 1, p. 19.
5. Bhutan, Country Economic Memorandum, May 1983, Vol. I, Main Report.
6. Ura, K., 1988, Economic Policy and Development in Bhutan, unpublished M.Phil Thesis, University of Edinburgh, pp. 47-48.
7. Seventh Five Year Plan 1992—1997, December 1991, Planning Commission, p. 141.
8. *Ibid.,* pp. 147-148.
9. *Druk Forestry News,* March 1993, Department of Forestry, Ministry of Agriculture, Royal Govt. of Bhutan, Vol. 1, No. 1.
10. Chhetri, D.B., December 1992, Forest Fire Season in Bhutan, in *Tsenden,* Vol. 4, No. 1.

7

INDUSTRY

— Bimalendu Bhattacharya

BHUTAN has a rich heritage of its traditional industries carried on for generations as examples of exceptional craftsmanship presented by works in bronze, silver and gold, wood, stone, papier mâché, clay, bamboo, cane, cotton, wool and silk. The wide array of objects made out of them ranging from statues of deities, bells, trumpets, swords, candlesticks, to doors and pillars of temples, tables, rice boxes, jewelry etc. As early as the sixteenth century a bronze foundry was in operation in the Punakha Dzong. Brightly painted, gilded sculptures of Lord Buddha and other saints are found in every temple. Ornamentally painted scrolls made on cotton and silk cloth adorn the walls of temples and buildings. Mythological figures made out of papier mâché, clay and wood are works of art commanding high skill. Handloom has a long tradition creating intricately patterned, choicefully coloured woollen and cotton fabrics. In the east, endi silk is spun from cocoons bred on castor oil plants. Furniture, bowls and dishes, and a large variety of items of daily use carved out of wood show exceptional craftsmanship. Articles like hats, baskets, butter containers, bows and arrows are made from finely worked bamboo and cane. Handmade paper of good quality in different shades of colour is produced from the bark of trees while gold and silver smithing, first introduced by the Newaris from Nepal still continues. All these traditional crafts are very much encouraged and supported by the government giving incentives for further development. A number of production centres have already come up at places like Thimphu, Paro, Dhokar, Tashigang and Tongsa providing training facilities along with credit and marketing assistance to the trainees on completion of their training period.

Industry, in a true sense, is, however, a recent development, making rapid strides from the early eighties though not yet taking a significant role in the national economy. The share of the industrial sector in GDP was only 4 per cent in 1980, rising to about 7 per cent in 1989. It remained underdeveloped even by the end of the Sixth Plan when several major projects were started, recording a high rate of growth of this sector.

In Bhutan, industries are classified into large, medium and small scale depending on the amount of investment. Factories having an investment exceeding Nu 1 million belong to the large category. Investments between Nu 15 lakhs and one million imply medium scale and those having less than Nu 15 lakhs as investment, fall into the category of small scale firms. "The classification is based purely on opening investment size of the firms. However, the distinction is not sustainable over time as some firms deteriorate in their capital stock over time if they do not replace them and others expand their capacity. Because of this, the firms which were initially identified as small in size may grow into medium size or *vice versa*" (Ura, 1988).[1]

A comparative picture of the development of industries between early and mid-eighties may give an idea about the progress made in this sector. For instance, in the early eighties there were about 70 industrial units of various sizes in the country developed in the private sector, 26 of them being wood-based firms. Besides, there were 27 saw mills spread all over the country. Among the important industries at this time were — a fruit preservation and processing unit, a steel furniture manufacturing unit, a gum resin factory, a match factory, a number of tea chest and packing case units. In the public sector, there were 3 distillaries, and a slate mining company. In late 1981, the first large industrial unit, the Penden Cement Plant with a 100,000 tonne capacity was commissioned in southern Bhutan which may be considered as a landmark in industrial progress in the country.

Right from the beginning the development of industries in modern line has largely been handicapped by 'lack of local technical enterprise and the overall shortage of unskilled labour'[2] in a country where the largest section of the working force was engaged in the agricultural sector. In addition to this great lacuna, a lack of infrastructural facilities making production costs much higher than usual, limited entrepreneurial enterprise and a limited local market were the other drawbacks at the initial stage of development.

However the enormous scope for development in this sector was not missed in view of Bhutan's enviable forest wealth, rich mineral resources, and the immense potential for agricultural diversification which together could provide a very strong base for the development of a wide range of industries with an eye to export market in the neighbouring countries.

The number of manufacturing plants stood at 349 in 1986, of which 335 were in the private sector and the remaining 14 in the public sector (Table 7.1). The overwhelming dominance of the food processing units comprising more than 71 per cent of the total number of industrial establishments — 249 out of a total of 349 — brings out the characterising elements of development in this sector. The manufacturing units of wood and paper products held second position with about 16 per cent of the total number of industrial plants in the country till then while the position of the other types was much less significant.

Table 7.1 : Number of Licensed Manufacturing and Mining Units in Bhutan, 1986

Industry	Public	Private	Total
Mining	4	5	9
Manufacturing :			
Food Processing	4	245	249
Textiles and Clothing	—	5	5
Wood and Paper Products	2	53	55
Chemical Products	1	11	12
Mineral Products	2	4	6
Other	1	12	3
Total Manufacturing	10	330	340
Total	14	335	349

Source : Statistical Year Book of Bhutan, 1990.

It appears from the Table that manufacturing developed till 1986 fell under five broad categories, they were :

(i) food processing/milling;

(ii) textiles and clothing;
(iii) wood and paper products;
(iv) chemical products; and
(v) mineral products.

Over and above this a number of different types of manufacturing activities lumped together, formed another category designated as 'other'.

The expansion of the food processing industry in the recent years is noteworthy. As pointed out earlier, growth of urban population and change in eating pattern are mainly responsible for the development of certain types of food industries, such as bakery and confectionery, concentrated mostly in the urban areas on the one hand while in another direction the food industry of late, has growingly become involved in giving rise to an altogether different set of industries like canning, bottling, oil and grain milling.

Bhutan Fruit Products in Samchi is a very large concern producing a wide range of items such as fruit juices, ketchup, jam, jelly, sauce and canned fruits. There are two small units producing fruit juices in Thimphu and Jakar in Bumthang. Lemon grass oil is produced for export to India for use in the manufacture of essential oils and perfumes. Mustard oil is produced for local consumption. There are three large distilleries at Samchi, Samdrupjongkhar and Geylegphug. Alcoholic drinks are made in Jakar. In 1988, the country exported alcoholic products comprising primarily rum and whisky worth Nu 18.4 million to India. Soft drinks are manufactured in Phuntsholing. The vegetable processing industry has not yet been able to make a strong foothold though dry (dehydrated) mushroom is exported in small quantity to Thailand and Japan fetching high prices. Among the other food industries sugar and milk products like butter and country cheese keep a low profile. Among the leading industrial units established during the Sixth Five Year Plan period are Dralha Flour Mill and Biscuit Manufacturing Unit in Phuntsholing. The Thimphu Agro-industry was to be completed at a cost of Nu 70 million.

The wood-based industries are mostly saw mills. There were altogether 49 of them in 1990 and all of them except two were small scale concerns operating on a temporary basis depending on the supply of logs. The machinery of these saw mills is inefficient but that keeps the cost of production low. Of the two large units, one, Gedu Manufacturing Corporation is an integrated establishment of saw mill, plywood mill and joinery factory. The large unit is Bhutan Particle Board Products Limited in Tala. In general, mechanisation is minimal in the saw mills which is the main reason for most of them do sawing on a contract basis for Bhutan Logging Corporation. The quality of sawing is generally not very good and major part of the secondary and tertiary production, such as joinery and furniture making are done by small concerns of which the latter — furniture making — represents the second largest wood-based industry in the country. Though not very developed yet, the types of wood-based industries operating in the country make an impressive list (Table 7.2), almost 75 per cent of them remaining concentrated in the western and southwestern parts of the country. Among the important plants set up during the Sixth Plan period are a Furniture Unit at Bhalujhora, and the Integrated Wood Complex at Lobesa. Gedu Wood Manufacturing Centre started production during this period. The Government also established a Furniture-cum-Training Unit in Thimphu at a cost of Nu 40 million at this time.

Table 7.2 : Wood Based Industries in Bhutan, 1990

Type	Number of Units
Sawmills	49
Sawmills-cum-Crate Boxes Plants	5
Particle Board Factory	1
Blackboard Factory	2
Plywood Factory	1
Joinery/Wood Works	4
Broomhandle Factory	3
Furniture Factory	33
Tea Chest/Crate Factory	4
Wooden Handicraft	1
Total	103

Source : Seventh Five Year Plan, Vol. I.

The major production of the Chemical Industry consists of carrier bags, polythene pipes, facial powder and lemon grass oil as a base for subsidiary industries.

The mineral-based industries, on the other hand, are gradually taking an important place in the national economy and some of them hold rich prospect in the perspective of industrial development in future. In fact, small scale mining work has been carried on for over a thousand years as it is revealed by the presence of slags in many parts of the country indicating the practice of mining lead-zinc and iron ores for the manufacture of artifacts, weapons, and most notably a number of iron-chain suspension bridges constructed in the 14th century by Saint Thangthong Gyalpo (1385–1464).

However, it is in the early 1960s that mineral exploitation on a modern scale started in Bhutan, and at present coal, limestone, dolomite, calc-tufa, slate, gypsum are mined for domestic industries as well as for export though their production constituted less than 1 per cent of GDP in 1990. Mineral production in Bhutan for 1990 is shown in Table 7.3.

Table 7.3 : Major Mining Companies in Bhutan in 1990

Minerals	Mining Company	Ownership	Output (MT)
Dolomite	Kalesore Dolomite Company Corporation, Private	Tashi Commercial	18,443
Dolomite	Khagrakhola Dolomite Mine	Chhundu, Private	19,708 (1989)
Dolomite	Bhutan Mining Enterprises	Private	17,971
Dolomite	Khanaburty Dolomite Mine	Singye, Private	34,400
Dolomite	Chainaikhola Dolomite Mine	Chaula , Private	114,050
Limestone	Yangzom Limestone Mine	Private	9,000
Limestone	Penden Limestone Mine	Government	138,324
Carbide	Bhutan Carbide and Chemical Limited	Joint	20,522 (1989)
Coal	Penden Duckpa Coal Company	Joint	20,000
Gypsum	Shumar Gypsum Mines	Government	40,000
Tufa	Kalapani Calca-Tufa Mine	Government	16,152
Slate	Shaa Slate Mine	Government	300,000 (sq ft, 1989)

Source : Seventh Five Year Plan, Vol. I.

The small amount of coal mined at Bhangtar in Samdrupjongkhar is exported to the neighbouring tea estates in India and Bangladesh. Exploration has revealed a total reserve of 89,000 tonnes of coal at 50 metres depth in Gerua block (Fig. 7.1). Dolomite which is of good quality and suitable for use in steel, glass and magnesium metal processing industries, is mined by five companies and exported to India and Bangladesh. The quartzite deposits at Tintale, Samchi, with a proved deposit estimated at 4 million tonnes have been found to be suitable for use by the Ferro Alloys Limited at Pasaka, a joint enterprise of the Government of Bhutan, a private Bhutanese company and a Japanese company. Another mine being opened at Chhukha will export quartzite to India and Bangladesh. The Ferro-Silicon Project at Pasaka came up during the Sixth Plan period.

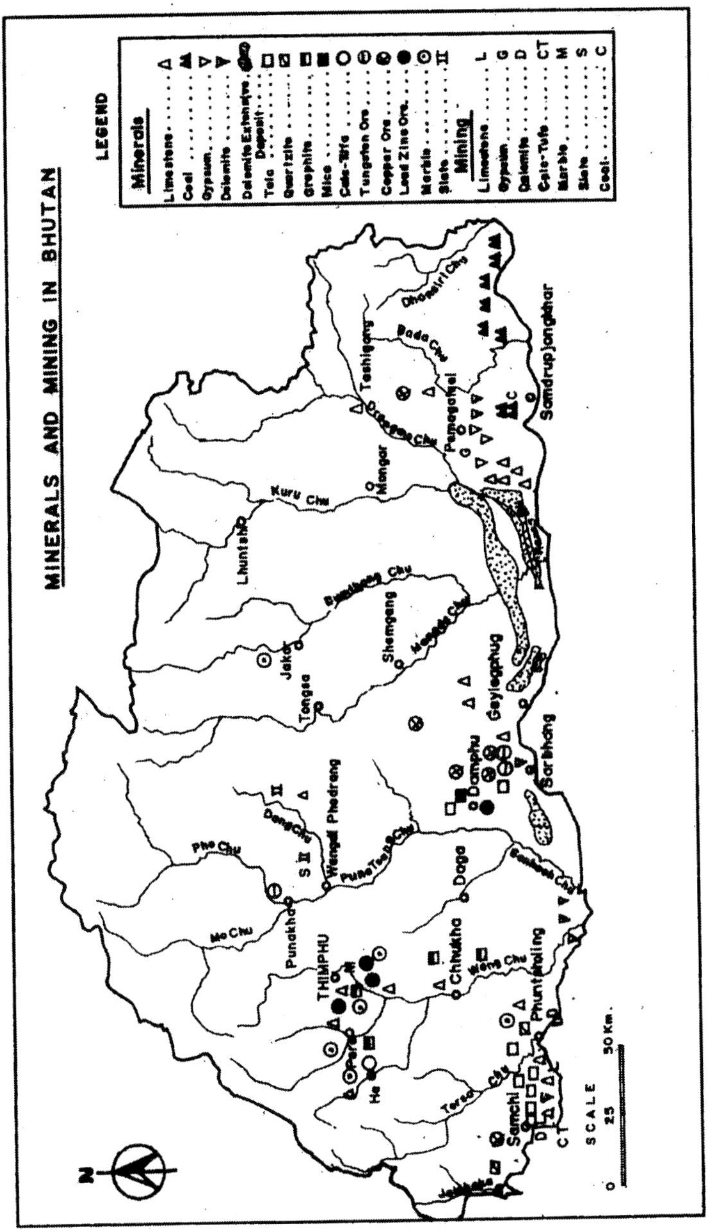

Fig. 7.1 : Minerals and Mining in Bhutan

Gypsum mined in Pemagatsel for use in cement plants is exported to India and Bangladesh. Limestone is mined for the manufacture of portland cement and calcium carbide. At present, there are three cement plants operating in Samchi district, each with a productive capacity of 400 tonnes, 100 tonnes and 30 tonnes respectively per day. The Penden Cement Authority, with a productive capacity of 400 tonnes per day is being supplied with limestone from Pugli Extension area, having a aproved reserve of 4.012 million tonnes. Among the most significant cases of industries set up during the Sixth Five Year Plan period are Yangzom Cement Plant at Bhalujhora, Lakhi Cement Plant at Gomtu, and the Dungsum Cement Project at Nanglam, the last of which will involve the establishment of a 1,500 TPD Dry Cement Plant with precalcinator. The requirement of about 100,000 metric tonnes of coal and 25,000 metric tonnes of gypsum will be met from local resources. The Gomtu Mini Cement Plant requiring 50,000 tonnes of limestone per year will have it from the reserve of 0.9 million tonnes at Titi area. The Marble Mining and Processing Unit at Gida Kom is an important addition during the period of the Sixth Plan. Bhutan Polythene Pipe Company and Bhutan Calcium Carbide Limited started production during the same period utilising the reserves of 1.94 million tonnes of cement grade limestone available in the Rongri area of Sarbhang.

There are altogether 33 mineral-based industries in the country (1990) majority of which (17) appear in the three western districts of Samchi, Chhukha and Paro, with the largest concentration of 11 in Samchi alone for the country as a whole. In terms of production, dolomite heads the list followed by limestone and these two among all the minerals in the country explored till now hold greater industrial prospect depending on, no doubt, a greater diversification of their use. The following Table 7.4 shows how the production of each of the four leading mined minerals has widely fluctuated over the years.

Table 7.4 : Estimated Production of Selected Minerals in Bhutan (in MT)

Minerals	1985	1986	1987	1988	1989
Dolomite	105,033	231,822	235,128	207,429	218,413
Limestone	161,596	192,424	177,706	168,704	131,546
Gypsum	10,198	24,838	20,146	27,997	48,308
Coal	26,669	30,155	36,600	19,066	20,816

Source : Seventh Five Year Plan, Vol. I.

Some important conclusions may be drawn from the character of ownership as well as the size-wise distribution of the mining and other industrial establishments in the country shown in Table 7.5.

Table 7.5 : Ownership and Size-wise Distribution (by district) of Mining and Industrial Establishments in Bhutan, November 1990

Dzongkhag OR District	Ownereship			Size			
	Private	Public	Joint	Small	Medium	Large	Cottage
Thimphu	196	6	5	140	5	3	59
Zone I							
Chhukha	97	2	3	62	11	7	22
Ha	21	—	—	15	—	—	6
Paro	45	1	—	33	1	—	12
Samchi	22	2	3	15	3	3	6
Zone II							
Chirang	12	—	—	7	—	—	5
Daga	—	—	—	—	—	—	—
Punakha	11	—	—	8	—	—	3
Wangdiphodrang	10	—	—	8	—	—	2
Zone III							
Bumthang	17	2	—	18	—	—	1
Geylegphug	44	3	—	22	1	1	23
Shemgang	9	—	—	5	—	—	4
Tongsa	5	—	—	1	1	—	3
Zone IV							
Lhuntshi	7	—	—	7	—	—	—
Mongar	27	—	—	25	—	—	2
Pemagatsel	5	—	—	3	—	—	2
Samdrupjongkhar	56	3	1	14	1	5	40
Tashigang	20	1	—	10	—	—	11
District not Identified	106	—	—	106	—	—	—
Total	710	20	12	499	23	19	201

Source : Statistical Year Book of Bhutan, 1990.

The most significant fact about ownership is that almost 96 per cent or 710 out of a total number of 742 of the manufacturing plants are in private ownership and not even 4 per cent is maintained by the government directly where joint ownership — private and public together — controls some of the plants. The public even joint ownership, is found to be highly restricted in its spatial occurrence, being absent in 11 out of 18 districts while one district, Thimphu, has the highest number — 11 out of a total of 32 industrial establishments — and the five western districts jointly have 21. The spatial distribution of the plants also brings out wide variations. There is at least one district, Daga, which has no industrial plant till the end of 1990 as the available data shows. On the other hand, Thimphu shows the highest concentration of industrial establishments, having nearly 28 per cent of the total, and more than 26 per cent are under private ownership. The second largest concentration of industrial plants occur in the four western districts of Chhukha (13.7 %), Paro (6.2 %), Samchi (3.6 %) and Ha (2.8 %). Thus, the five western districts together contain more than 54 per cent of the total number of industrial establishments in the country. Three other districts, namely Samdrupjongkhar (8.1 %), Geylegphug (6.1 %) and Mongar (3.6 %) in the east have a fairly large number of industrial establishments, together having 17.8 per cent or 127 out of a total of 742 in Bhutan.

The disparity in the distribution of industrial establishments of different size-categories is also equally noteworthy. The small scale units comprise over 92 per cent of the total and except one district — Daga — they are found in all the districts. There are 11 districts where both medium and large scale establishments are absent. Again, majority of the industrial establishments have developed in a concentrated manner in the same five western districts mentioned earlier with over 55 per cent of the total, and among them Thimphu alone with over 27 per cent belonging to different size-categories leads. Chhukha with 7 large scale industries heads the list followed by Samdrupjongkhar with 5, while Thimphu lags behind in this respect having only 3 large scale units. The most significant change in industrial development in the country in recent years is revealed in the fact that the number of cottage industrial units has widely been surpassed by the modern industrial set-ups, the total number of the former being only 201 in spite of encouragement and incentives provided by the government. Apart from three districts — Ha, Daga and Lhuntshi — units of cottage industry are found in all others with major concentration in the four western districts of Thimphu, Chhukha, Paro and Samchi, containing almost 49 per cent of the total. Thimphu with 29 per cent of the cottage industrial units leads and Samdrupjongkhar in the extreme east having 20 per cent of them has the second largest concentration in the country.

One of the major objectives for the Seventh Five Year Plan (1992-97) was to 'increase the share of the manufacturing sector from 7 per cent in 1989 to 13 per cent by 1997 at an average growth rate of 13.4 per cent per year'. Another major objective was to 'substitute imports through domestic manufacturing'. It was felt that greater economic efficiency in the manufacturing sector could be achieved by further 'privatisation of public enterprises and contracting out of many of the departmental activities to the private sector'. In order 'to promote more regionally balanced growth in manufacturing, incentive packages were to be provided for industries for their development in less developed areas'. For this purpose, priorities have been determined for the districts on the basis of the stage of their industrial development making first, second and third priority districts and industries set up in the first were to get 20 per cent of the project cost as subsidy, second and third priorities respectively 10 and 5 per cent. In this connection both identification and feasibility studies of the projects were to be taken up and further assistance will be provided to the private sector through specialists in the field.

Production and distillation plants for the extraction of essential oils have been identified as projects with great potential, particularly in eastern Bhutan where the oil plants like palma rosa and lemon grass are used for this purpose. A brewery with a production capacity of 50,000 hectolitres has been proposed using barley which can be grown in many parts of the country and 70 per cent of the beer produced is meant for export to the neighbouring countries. A readymade garment Manufacturing Unit will be established for producing dresses and garments for export to America and Europe. A Particle Board Unit will be set up with an installed capacity of 15,000 cubic metres per year for export to India and Bangladesh.

Major strides are also intended to be taken under the Seventh Five Year Plan. The Ferro Silicon Project under implementation is a major private sector enterprise with output capacity of 15,000 tonnes per year of ferro silicon, an important ingredient in the production of steel. A training centre will be opened at Gida Kom for improving the quality and increasing the production of the wood industries. In order to provide necessary infrastructural facilities for the private sector enterprises, 12 industrial estates and 12 service stations will be established , making provisions for power, water and extension services.

The Seventh Plan has laid greater emphasis on two major areas in the perspective of mineral development. One of them is detailed geological mapping, and the other is mineral exploration. These included programmes on mapping of the western extension, such as, Bhurkhola for tungsten; Wangdiphodrang for construction materials; Urichu and Khar for gypsum; Kalikhola for phosphate; Bhurichu and Chirang for quartzite; and Gurungkhola, Thimphu, Ha and Paro for zinc and lead. At the same time appraisal of the mineral deposits will be carried out for iron ore deposits at Maure and Kalikhola; for talc at Samchi; for mica at Chirang. Along with this exploration and investigation will be made for tungsten at Nobding, for tertiary granite at Wangdiphodrang and Chirang, and for pegmatite bodies at Merak Sakten and Bidung.

REFERENCES

1. Ura, K., 1988, *Economic Policy and Development in Bhutan*, p. 72.
2. Bhutan, Country Economic Memorandum, May 1983, p 26.
3. Seventh Five Year Plan 1992—1997, December 1991, Planning Commission.

8

TRANSPORT AND TOURISM

— *Bimalendu Bhattacharya*

Transport

IN a mountainous country like Bhutan roads happen to be the only means of surface transport but their construction and maintenance are far from easy tasks to provide wide accessibility. The difficult terrain, far-flung situation or dispersed character of habitations, varied and severe climate coupled with an overall dearth of trained personnel and labour force have made road construction a very expensive activity. The geographical location of Bhutan between Tibet in the north and India in the south has naturally made it a corridor for trade and cultural contacts between the two outlying countries for which trails criss-crossing the country for the use of man and animals have long provided the only means of contact. The extensive trade relations had developed through these routes since ancient times and was carried on till recently with Tibet via Chumbi Valley in that country establishing direct links between Lhasa, Paro, Punakha and Tashigang. The route from Paro was through Tremo La to Phari; from Punakha it used to follow the Mo Chu valley for going to Gyangtse and Shigatse in Tibet, and from Tashigang the route lay via Dozam valley to Shingbe. From eastern Bhutan the trade followed the course of Lhobrak and Dozam rivers to Tibet. Yatung in Tibet was the entrepot of commerce and the main centre of exchange for caravans arriving from Paro, Thimphu and Punakha.

But alignments of the high mountains lying across the major routes along with unfordable, torrential courses of the upper reaches of the rivers have always been the main obstacle to the development of these routes, giving rise to long detours from which there is no escape even today. While the mountain passes provided passages at many instances, the construction of suspension bridges at strategic points have long been a major step in this direction to get over this impediment the major instances of which may be traced to as early as 14th century. However, development of good roads did not find favour with the government till early 1950s due to "conscious decision not to undertake any road construction"[1], and thereby keep "Bhutan largely inaccessible to outsiders. Thus, Bhutan turned down several offers of roads, including those early in the century by the British who had been anxious to develop more convenient links to Tibet".[2]

The history of road development in modern sense made a new chapter with a beginning in the late 1950s which synchronised with the decision of the economic development of the country. In order to establish closer physical links with the outside world, the Bhutan Engineering Service (BES) started in 1959 the construction of the first motorable road linking up Phuntsholing with Thimphu (179 km) and Paro using compulsory labour. But it would have really been a very difficult task for Bhutan to go alone in its venture on road construction drawing labour away from agriculture for a long time if the assistance was not obtained from the Indian Border Road Organisation (IBRO) in 1959 bringing with it a large force of contract labour. On completion of this road in 1962, the construction of the second road, a north-south link road between Samdrupjongkhar and Tashigang (180 km) in the east was taken up in 1963 and completed with the assistance of the IBRO. At the same time the road to Paro was widened and surfaced.

Road construction came to be treated as a high priority area in all the Five Year Plans since their inception in 1961, when a Public Works Department was set up for all construction works including roads and bridges and 58 per cent of the total outlay in the First Five Year Plan was set aside for road construction. During the Second Plan period which had one-third of its total outlay involved in road construction, the IBRO continued to broaden and surface these roads while a third link was established connecting the southern border with the interior from Geylegphug to Tongsa (244 km) in central Bhutan. Further, the construction of an east-west link road connecting Thimphu and Paro with Tashigang (546 km) in the east started by the IBRO in 1965 was completed and opened for through traffic in 1975 when 200 km of the 546 km road was already paved. By this time the development of a basic road network was nearly complete and the emphasis was shifted to roads linking the district headquarters to the main network and to feeder roads. In the Third Five Year Plan road construction and maintenance enjoyed the same priority though expenditure on road development was slightly reduced (17.8%). Even in the Fourth Plan its share was next only to agriculture and education, and by late 1982, there were 2,050 kilometres of roads in Bhutan, including National Highways, district roads and feeder roads connecting all but two district headquarters with the major part of the country and of the total length of roads, about 1,200 km were black-topped all-weather roads and the rest earthen. By this time there were three major north-south black-topped highways traversing the country respectively in western, central and eastern Bhutan and one fair weather lateral road running from east to west. The phenomenal development in road construction primarily aimed at providing access to 18 district headquarters connecting them with 12 sub-divisional (Dungkhags) headquarters, providing 0.044 km per sq km of level area and about 1.77 km per 1,000 population.[3]

In spite of these efforts lack of a well developed transport network has still remained as one of the major constraints to the development of the country where even now the majority of the people live at least half a day's walking distance from the motor road and depend entirely on mule tracks and foot trails connected by suspension and cantilever bridges and in a few places by ropeways to get access to the nearest large settlements.

Accessibility has remained poor till now and the least accessible areas are in the south-eastern parts of Bhutan, comprising particularly Kheng and some parts of Mongar, Samdrupjongkhar, Pemagatsel and Tashigang districts. The road density is extremely poor in these areas, often requiring 3 to 6 days of trekking to the roadhead.

Hence, the importance of foot and mule tracks was also given due recognition along with the development of paved and unpaved roads and elaborate programmes were taken up and executed for their improvement including foot suspension bridges. Eighty such bridges have been built till the end of the Fourth Plan to break the isolation of the remote areas suffering long from poor accessibility.

The Fifth Plan (1981—87)[4] included a number of important targets. It was planned to make some new roads and a network of feeder roads.

The new roads connecting the district centres to the existing National Highways included a road between Wangdiphodrang and Chirang, another to connect Gedu with the forest areas in the interior and the third was from Samchi to Ha, Paro and Thimphu. The black-topping of the East-West Highway between Wangdiphodrang and Tashigang was also included in the programme. By this time it was estimated that the overall requirement of mule track to connect 4,500 settlements in the country amounted to about 2,500 km and about 1,000 km of the existing mule-tracks were to be improved, upgraded and extended during the Fifth Plan period while the task of planning, execution and maintenance of mule-tracks was entrusted with the district administration.

A Review of Road Development between Fourth and Sixth Plan Period

The progress in road construction seems to have been much less impressive during the period of the Sixth Plan than earlier due to heavy monsoon in the first three years besides such factors stalling progress being acute shortage of skilled and technical manpower, and late arrival of necessary equipment. The slowing down in the progress is quite evident from the data available till 1990, the completion of the third year of the Sixth Plan (Table 8.1).

Table 8.1 : Cumulated Road Length in Bhutan constructed since the Fourth Five Year Plan (km)

Category	Fourth Plan	Fifth Plan	Sixth Plan (till 1990)
National Highways	1,358	1,450	1,453
District Roads	216	447	500
Feeder Roads	185	268	408
Total	1,759	2,168	2,361

Source : Seventh Five Year Plan Vol. I.

Table 8.1 shows that there had hardly been any progress in the construction of the National Highways during the first three years of the Sixth Plan and this is also partly true about the District Roads of which 231 km was constructed during the Fifth Plan period as against 53 km during the three years of the Sixth Plan. The only compensating factor is about the feeder roads having increased their length by 140 km during this period as compared with increase in their length (83 km) during the entire period of the Fifth Plan.

The picture appears to be equally dismal as regards achieving the target in the construction of 120 suspension bridges during the Sixth Plan period 31 of which were completed by the end of 1990 and work on 15 others was in progress against which a total of 122 were completed in the Fifth Plan and 90 in the Fourth Plan.

Unlike the earlier Plans, the Seventh Five Year Plan put greater emphasis on the repair and maintenance of the road network, particularly the highways, adducing the highest priority to this than new constructions so that the expansion of the road network does not exceed maintenance capacity. Hence, rather than setting any new target for the construction of new roads, its objective was to take up the task of completing the spillover works. However, construction of new roads was to be taken up only in the southeastern region of the country, particularly in Pemagatsel and Tashigang districts.

One of the important strategies of the Seventh Five Year Plan was to involve the private sector in the construction and maintenance of roads on a contract basis. The response has been found to be encouraging. Another major strategy as adopted in the Sixth Plan, such as mechanisation of road construction and maintenance was to continue. During this period a total backlog of about 710 km, comprising both national highways (580 km) and district roads (130 km) were to be surfaced. Besides, 75 bridges were to be built which included bridges with steel towers, suspended bridges, reinforced concrete bridges, wooden beam bridges, and traditional wooden cantilever bridges in addition to improvement of the existing bridges.

Airways

The development of airways is still at its incipient stage handicapped by many factors, the very limited clientele being one of them. The national airline, the Druk Air Corporation, a government undertaking, was formed in 1981 beginning its operation in 1983 with three flights a week from Paro where an airfield existed already. Initially, an 18 seater Dornier aircraft was used for this route to which another Dornier was added later on. Recently, a new aircraft—BAe-146—with a seating capacity of 88 passengers was purchased which increased the air-carrying capacity appreciably. However, air service to Bhutan is still limited to Paro alone though there are many helipads in the country. Besides Calcutta with which Bhutan maintains air service twice a week, air links have been established with Dhaka in Bangladesh in 1986, with New Delhi in November, 1988, with Kathmandu in Nepal in April, 1989, and with Bangkok in March, 1989. "The Civil Aviation provides infrastructure at Paro Airport, air traffic control facilities, navigational and communication facilities and security for National Airline. It is responsible for the formulation of rules and regulations to ensure safe operation of Royal Bhutan Airlines and enters into bilateral agreements with other countries to facilitate the establishment of air links."[5] The main objective of the government to make the Druk Air Corporation commercially viable has not yet been achieved which depends on the increase in the number of overseas tourists, and in spite of raising the ceiling for them to 4,000 a year, the country has not so far been able to attract that number in any year.

On the whole, however, the air traffic is on the increase as is evident from the flow of air passenger handled by the Druk Air, recording an appreciable rise of 38.8 per cent in one year between 1988-89 and 1989-90 (Table 8.2). There are, however, some interesting points to be noted in the number of passengers travelling different routes. For instance, in 1988-89 the component of passengers between Paro and Calcutta was the largest,

Table 8.2 : Passenger Traffic by Druk Air

Stations	June 1988—June 1989			June 1989—June 1990		
	Arrival	Departure	Total	Arrival	Departure	Total
Paro-Delhi	1,391	1,558	2,949	3,117	3,113	6,230
Paro-Bangkok	—	—	627	1,355	1,620	2,975
Paro-Dhaka	386	420	806	316	406	722
Paro-Kathmandu	249	226	475	889	903	1,792
Paro-Calcutta	3,193	3,035	6,228	1,983	1,688	3,671
Total			11,085			15,390

Source : Seventh Five Year Plan, Vol. I.

representing over 56 per cent of the total air traffic followed by those taking Paro-Delhi route with over 26 per cent. The remaining, a little over 17 per cent took the other three routes, Paro-Dhaka (7.3 %), Paro-Bangkok (5.6 %) and Paro-Kathmandu (4.3 %). The picture reverses in the following year (1989-90), the Paro-Delhi route carrying the largest volume

(40.5 %), followed by Paro-Calcutta (23.8 %) and Paro-Bangkok (19.3 %). Paro-Calcutta's share declined by more than 41 per cent while that of Paro-Delhi route increased by nearly 53 per cent and the share of Paro-Kathmandu route recorded a gigantic leap, increasing by over 277 per cent. The marked decline in passenger volume on Paro-Calcutta route is somehow compensated by considerable increase on other routes except on Paro-Dhaka route which fell by about 10 per cent.

Tourism

Tucked away in the interior of the Eastern Himalaya, Bhutan has a unique cultural heritage strongly manifest in its art, music, literature, traditional crafts and architecture developed in a widely varied physical setting which has a strong attraction for people coming from other countries. In that sense the country holds a high prospect for tourism and, given due importance with proper attention, it may develop as a major source of earning foreign exchange. The real potential of tourism has not escaped the attention of the government which is interested in utilising this resource for economic benefit without impairing the cultural background of the country. The approach in this regard has therefore been in the form of promoting package tours particularly for foreign tourists fulfilling a wide range of interests so that exposure of the country can be kept under strong control.

Accordingly, the first batch of tourists visited Bhutan in 1974 and since then the total number increased to nearly 2,000 a year by 1980-81, contributing 1.3 million U.S. dollars or Nu 11.7 million in 1981-82 to the national exchequer.[6] By the end of the Fourth Plan, 5 hotels with about 200 beds came up at important locations in the country and a Travel Agency under the Departnent of Tourism has also been establised to maintain contact with foreign agencies for the promotion of tourism.

Tourism had so far been limited to sight-seeing trips to the western part of the country in the absence of infrastructural facilities and this prompted the planners to accord high priority in the Fifth Plan to the development of specific resorts for this purpose in the other parts of Bhutan which were to be linked up with such activities as trekking. Touring of cultural areas like dzongs and monasteries were to get the highest priority followed by fishing as a game for which suitable areas were to be developed. One of the major objectives of the Sixth Plan was to make greater efforts to earn foreign exchange from tourism while protecting the environment and culture from possible adverse consequences of tourism. However, the number of tourists did not increase immediately, stagnating for a while in the lack of necessary facilities needed to be expanded and improved with greater stress on publicity. The situation improved much with the privatisation of the Bhutan Tourism Corporation in late 1991 since when several travel agencies came to be involved and the major policy guidelines and regulations on tourism were framed by the Board of Tourism. The Board decided to allow upto 4,000 tourists a year during the Seventh Plan period (1992—97), the increased inflow to be made possible through the expansion of tourist facilities.

The information available for 1994 and 1995 reflect the results of improved infrastructure, and if this trend is maintained, Bhutan is likely to derive much greater benefit from tourist trade. It appears that the total number of tourist arrival in 1994 was 3,971 and it was 1,617 for the first four months of 1995 (according to available data) which, in fact, is more than 35 per cent higher for the same period in the previous year. The total earnings on this account amounted to nearly 4 million U.S. dollars (Table 8.3) and over 1.7 million U.S. dollars for the first four months of 1995, being a little over 0.6 million U.S. dollars than for the corresponding period

Table 8.3 : Tourist Arrivals in Bhutan and International Tourism Receipts

Months Arrivals	International Tourist		International Tourism Receipts (million U.S. dollars)	
	1994	1995	1994	1995
January	101	120	0.058	0.066
February	175	134	0.103	0.088
March	643	341	0.695	0.418
April	274	1,022	0.295	1.187
May	373		0.287	
June	115		0.079	
July	99		0.088	
August	301		0.235	
September	465		0.546	
October	787		1.038	
November	418		0.410	
December	220		0.137	
Total	3,971		3.968	1.759

Source : Bhutan Tourism Corporation, Thimphu, Bhutan.

of the previous year. The peak season for tourism seems to be from August to November when almost half (49.63 %) of the tourists arrive followed by another peak between March and May having about one-third (32.48 %) of the total arrivals. In 1995, more than three-fourths (84.29 %) of the total number of tourists for the first four months of that year came in March and April which presumably shows the seasonal trend where the winter months are less favoured by the tourists.

It may be of interest to look at the country of origin of the tourists to ascertain the degree of internationality. The data available on this account from Bhutan Tourism Corporation giving a countrywise break-up shows 16 countries as specific points of origin and two more as groups of countries, like 'Other Europe' and 'Other World' (Table 8.4) both for the year 1994 and 1995 (the first four months). The largest number of arrivals, comprising 25.9 per cent was from Japan in 1994, putting the country at the head of the list as well as for 1995 with 27.9 per cent of the total number of arrivals

Table 8.4 : Foreign Tourists in Bhutan by Country of Origin

Country of Origin	Arrivals			
	1994		1995 (first four months)	
	Number	in %	Number	in %
1. Japan	1,029	25.9	451	27.9
2. United States	689	17.4	234	14.5
3. Germany	566	14.2	204	12.6
4. United Kingdom	304	7.7	82	5.1
5. France	219	5.5	121	7.5
6. Italy	202	5.1	103	6.4
7. Netherlands	165	4.2	37	2.3
8. O.C. Asia	131	3.3	35	2.2
9. Other World	121	3.0	54	3.3
10. Switzerland	116	2.9	36	2.1
11. Australia	91	2.3	54	2.3
12. Belgium	85	2.1	48	3.0
13. Austria	80	2.0	66	4.1
14. Other Europe	52	1.3	34	2.1
15. Canada	45	1.1	0	-
16. Spain	44	1.1	9	0.6
17. Sweden	26	0.7	23	1.4
18. New Zealand	6	0.1	26	1.6
Total	3,971	100.0	1,617	100.0

Source : Bhutan Tourism Corporation, Thimphu, Bhutan.

in the first four months. Thus, over one-fourth of the foreign tourists in both the years was from Japan followed by those coming from the United States and Germany in both the years. However, on a close scrutiny Europe is ahead of all countries, contributing 46.81 per cent of the total number of tourists in 1994 and surprisingly enough, representing almost the same proportion (46.62 %) of the total arrivals from abroad in the first four months of 1995. Even taken separately, four countries of Europe, namely Germany, the United Kingdom, France and Italy keep between them third to sixth places in the total number of tourists in both the successive years. In short, the character of tourists in terms of nationality is more European till now than otherwise though data not even for two whole years can be adequate to be treated as trend-setter but the international character of the tourists with origin from a quite a large number of countries of the world is firmly established.

REFERENCES

1. A World Bank Country Study : Bhutan, Development in a Himalayan Kingdom, 1984, Washington, p. 71.
2. *Op. cit.*
3. *Op. cit.*
4. Fifth Five Year Plan 1981-1987, p. 37.
5. Seventh Five Year Plan 1992-1997, December 1991, Planning Commission, p. 116.
6. Bhutan, Country Economic Memorandum, May 1983, p. 29.

9

DEVELOPMENT STRATEGY

— *Neil Fraser*

Economy Before Modernisation

Before the 1950s Bhutan had a customary self-subsistence economy with poor internal communications and external isolation. There was a lot of exchange within valley communities, but only a little over the Himalayas to Tibet or south to India. Farming systems generally met subsistence needs, using mixed farming methods with little specialisation, which were well adapted to the risk-prone environment and climate, and local arrangements for use of collective resources. Local elites would collect taxes-in-hand, labour contributions and militia duties, whilst some families would be in effect their serfs. The monasteries also required contributions. There seems to have been generally enough food for all and timber and stone for housing for all. But morbidity and mortality were high, and literacy low. The local autonomy of villages was largely undisturbed by central institutions, the monarch and his regional representatives (Ura,1994).[1] What trade, commerce and industry there was, was left by the farmers to the royal family.

The Beginnings of Modernisation under the Third King

Under the first two kings Bhutan maintained its isolation, with some political centralisation but little economic change. But the third king, Jigme Dorji Wangchuk (1952---72) set the country on a path of modernisation. Rose discusses the factors behind this. One was the new king's background, including some schooling in India. Secondly, Bhutan could not remain isolated from external influences with the threat seen to lie in aggressive Chinese policies in Tibet and India's anxiety to block the Chinese. Thirdly, the king recognised the dangers in not seeking some political changes in Bhutan, and the opportunities which some 'development' could bring e.g. expanding government revenues and getting new sources of employment and production. What is more India was willing to pay for the development moves once the Chinese threat was recognised as real.

The king's first reforms were (1) a land reform programme and (2) the setting up of a National Assembly (the Tshogdu). The land reform ended the remaining feudal tenancy system, by allowing tenants/serfs to acquire their own land — an important step, though by the 1950s it affected relatively few households. There was also an attempt to establish land ownership ceilings, but (as in other countries) this was little implemented. The Tshogdu was a potentially radical innovation for an absolute monarch. Membership consists of three groups : certain office holders like ministers and district secretaries, representatives of the monk body, and representatives of the people. The king and the Tshogdu are described as 'dual sovereign powers' but in practice it has been a conservative body, dominated by officials and local leaders, largely focused on local subjects. A subject which had to be included in any modernisation programme was education – its extension beyond religious schools to provide the personnel needed for economic development. Rose (1978) argues that the king was willing to postpone modernising the economy until there were enough skilled Bhutanese, but this cautious approach was abandoned in 1960 when the decision was taken to respond to the Chinese policy in Tibet by road-building to link Bhutan strategically with India.[2] Nehru had offered Bhutan aid in 1958 but Bhutan was reluctant until the escape of the Dalai Lama in 1959.

Acceptance of road-building and military training from India led Bhutan into the approach of having five-year plans. The first plan, in 1961, was mainly a statement of projects India was willing to support, and was 100 per cent funded by India. Nearly 60 per cent was expenditure on road works. It was a fateful step. Bhutan was placed clearly in the Indian sphere of influence, with road-building promising trade and development in collaboration with India (in 1960 Bhutan actually banned trade with Tibet/China — several years before the road was completed giving the link with India).

Quite apart from the strategic arguments for roads, road building was a natural first step in economic development. The country have little trade, internal or external, as long as communications were limited to horseback along mountain trails. Roads were also a pre-condition of central, as opposed to local, action for 'basic needs' development e.g. for general education and health. The early road-building was not only paid for by India, but was accomplished by Indian engineers and labourers. External dependence versus self-reliance has been a major issue for Bhutan's development since these early days.

The Second Plan (1966-71) was, like the first, on a small scale, Nu 202 million, and almost entirely funded by India. Again spending on road-building was a large part, but there was the beginning of project spending on education, health and agriculture. The aim of education in the early plans was to provide skilled manpower for development – starting from a very low base, and therefore heavy dependency on Indian personnel.

Rose (1978) argues that a number of the early moves in Bhutan's development helped to pacify the peasantry. One was the land reform in 1952 already referred to . A second was the substitution of cash land taxes for in kind taxes in 1970 and their reduction to low levels. This meant that, unlike in some countries, the farmers were not being asked to pay for the development. A third was some reform of the compulsory labour contribution (*chunidom*) from each household, although it was not abolished. Rose also argues (1994) that there was surprisingly little protest from local elites about officials of the new ministries concerned with development usurping their powers.[3] The existence of the Tshogdu, which seemed to give these elites a voice, may be one reason. There was later a decentralisation policy launched by the king (see below).

Developing the Development Strategy

It was not until around 1968 that there was a general development of ministries in Bhutan. The first administration separate from the Royal Palace was the Development Secretariat preparing the five year plans, working with a lot of Indian staff. From the start of the Third Plan (1971) this became the Planning Commission. This plan, which spent Nu 475 million, was still nearly 90 per cent funded by India, with 6.9 per cent from internal sources and the remainder from international organisations. One of the tasks of the new Planning Commission was to diversify sources of aid beyond India.

But there was also a question of overdependence on aid itself. In the Fourth Plan (1976), for instance, which spent Nu 1,106 million, 77 per cent was funded by India, but only 5.4 per cent was funded internally. The first four plans, as Rose (1978) has remarked, reflected more what India was willing to provide rather than Bhutan Government policies.

I want first to address the question of the direction of development. This should be considered in relation to the resources and constraints for the country. Bhutan is rich in certain resources: hydro-electric potential, forests, certain minerals like limestone. But it is extremely short of capital, has labour shortage with a mainly rural population lacking 'modern' skills, has no extra arable land to cultivate, has a small domestic market and little experience of entrepreneurship. And even with some roads, trade is constrained by the mountainous environment.

These constraints, as Shah (1989) has argued[4], ruled out broad-based approaches like 'liberalising' trade and investment, or import-substituting industrialisation. But there would still seem to be a big choice between a 'rural development/basic needs' approach and the selective economic development of key resources. In the case of Bhutan the initial emphasis was on 'resource-based growth' especially power projects, and industries using power, mining, and forestry, and to a lesser extent cash crops and agro-industries.This approach promises economic growth, including State revenues which could be used for social development and reduced dependence on foreign aid. Rural development, including help for agriculture, providing schools, health clinics and other rural amenities, and seeking to improve the quality of life of ordinary people, was not so evident in early plans, but has come more into prominence from the Fifth Plan onwards. A difficult question to evaluate is whether general incomes and living standards might have been higher with any different strategy.

India has provided most of its aid on very favourable terms, without specifying its use. It has been available for Bhutan's current trade deficit as well as for development projects. They have also funded (in part) projects outside the five year plans, notably the biggest of the resource-based projects, the Chhukha Hydro-Electric Scheme.

The objectives of the three recent Plans, numbers Five to Seven, have a lot in common. In contrast to the first four Plans, the Bhutan Government began to seek what it called 'self-reliance'. This involved a range of sub-objectives like revenue generation, decentralisation, human resource development, and conservation of the environment and the country's traditions. The objective of 'self-reliance' will be discussed in a section below. Another objective has been the strengthening of the public sector. In terms of sectors these plans have involved both economic (manufacturing, power, telecommunications) and social sectors (education, health). Agricultural spending can be seen as both economic and social – there is a continuing emphasis on productivity improvement but it is within a basically traditional framework of family farms. Table 9.1 below summarises the sectoral breakdown.

Table 9.1 : Percentage Expenditure on different Categories in the Fifth, Sixth and Seventh Plans

Heads	Fifth Plan	Sixth Plan	Seventh Plan
Agriculture	12.5	12.7	12.6
Forests	4.9	4.4	3.1
Social Services	20.3	14.9	22.7
Power	7.3	13.1	2.5
Trade and industry	7.0	13.3	6.5
Roads/public work	16.9	9.3	7.8
Telecommunications	0.7	1.0	7.0
Other	30.4	30.9	37.8

Note 1 : Much expenditure, particularly for power and industry, is outside the Plans.
2 : Categories may have boundary changes between Plans.

Sources : *Statistical Year Book of Bhutan* 1990, Table 13.2; Seventh Five Year Plan, Vol. 1, Table 7.2

Choice of Key Resource-based Projects

Chhukha Hydro-Electric Scheme

This scheme, with a capacity of 325 MW, was completed in 1988, at a cost of Nu 2.44 billion, made available by India as 60 per cent grant and 40 per cent loan. Total revenue, mainly from sale of power to India, in early years was Nu 380 million per annum and net revenue to the Bhutan Government Nu 240 million per annum. Although it contributes 16 per cent to the Bhutan Government revenue (92/3) it can be said to be sold at a very uncommercial price (see World Bank, 1994[5]). But Bhutan has to accept that because the funding was very uncommercial, India agreed to buy all power not wanted by Bhutan at an agreed price. Around 90 per cent of the power has been bought by India. It also raised the chance of power-intensive industries being profitable for Bhutan.

Manufacturing Industries

Penden Cement, which began operation in 1981, was the first big industrial project in the strategy. It was developed in time to supply a lot of cement for the Chhukha dam. Its output in 1993 was Nu 114 million of which 98 million was exported.

Two sizable wood product companies are Gedu Wood Manufacturing (established 1987) and Bhutan Board Products Ltd. (established 1990). Their 1993 output was Nu 67.6 million and Nu 165.9 million respectively. However, the exploitation of forests by Bhutan is more notable for caution on environmental grounds than the pursuit of income.

A later industry which is very energy intensive (taking at one time 80 per cent of Bhutan's industrial consumption of electricity) is the manufacture of calcium carbide, which is used in the production of acetylene. This began in 1989 and is almost entirely exported to India. Bhutan Carbide and Chemicals Ltd. had an output of Nu 357 million in 1993.

A ferro-alloy plant at Pasakha came into production in 1995. Again it is very energy intensive. It is the first large scale industry to be set up in Bhutan with foreign investment — from Japan (Kuensel, April 1995)[6]. Ferro-silicon is exported for use in the manufacture of steel.

Of export-based agro-industries the most important is Bhutan Fruit Products Ltd., with 1993 sales of Nu 92 million, made up of canned fruit, juices, jams etc.

Relatively big firms, mostly created by the public sector, form the mainstay of Bhutan's industry. In 1986, four firms produced 80 per cent of Bhutan's recorded industrial output (producing cement, fruit, wood products and alcoholic drinks).

The share of industries and power in the development plans is shown in Table 9.1 but the figures can be misleading as these sectors involve large non-plan expenditure (e.g. Chhukha cost the equivalent of 52 per cent of the total Fifth Plan).

Plate 13 : Chhukha Hydel Project

Plate 14 : An Eye Chorten in Central Bhutan

The estimated share of GDP from manufacturing is 3.2 per cent in 1980 and 6.0 per cent in 1989 (CSO, in Seventh Plan). The impact of the strategy on internal revenue generation is indicated by the growth in revenue/GDP ratio for Bhutan from 10.9 per cent in 1983-84 to around 20.0 per cent in 1988-89, 1989-90, 1992-93 and 1993-94. Revenues exceed grant aid after 1988. The impact of the strategy on exports can be judged by the growth of exports from Nu 160 million in 1983 to Nu 1,042 million in 1988. However, exports were then no higher in dollar terms for the next four years (IMF, quoted by World Bank, 1994). Electricity made up 26 per cent of merchandise exports in 1993 (World Bank, 1994).

Rural Development/Basic Needs

It can be argued that rural development was a neglected aspect of early development programmes in Bhutan. Education spending was primarily geared to skill needs (especially for the Civil Service) and reforming traditional agriculture was not an emphasis. This has changed, arguably under the influence of the decentralisation policies. Social services were planned to take up 22 per cent of public expenditure in the Seventh Plan period compared with 15 per cent in the Sixth Plan period. Education and health services are discussed further in a separate chapter.

Village level services are hampered by major obstacles. They need a lot of skilled staff — teachers, doctors, agricultural extension officers, water engineers — of which Bhutan is very short. And the sparse, scattered population in the mountainous terrain, with few roads, makes provision of services very difficult and costly. A further important point is that capital expenditure in these areas leads to high recurrent costs. Aid funding is more often available for capital costs than for the recurrent costs in perpetuum thereafter.

Social Indicators play an important part in motivating basic needs programmes (once a country is concerned about its international image). Bhutan's development began from a very low base, judging by early indicators (themselves based on small samples and poor demographic data) (Table 9.2). Some examples :

Table 9.2 : Social Indicators for Bhutan

Life Expectancy	48 years (1991)	66 years (1994)**
Infant Mortality rate per 1,000 live births	182 (1970)	132 (1991)
Adult Illiteracy		62% (1990)
Percentage of Age Group enrolled in Primary education	6% (1970)	26% (1990)
Daily calorie supply per capita	2,477*	

Source : World Bank 1993, World Development Report, (*1988 Report)
World Bank 1994, Bhutan : Country Economic Memorandum **

The nutrition indicator is good by the standards of low income countries, but not mortality and literacy. The widespread ownership of land, and not much evident poverty, perhaps reduced pressure to provide basic needs programmes (Shah, 1989).[7] But that people do want schooling and health centres became clear over time and fuelled pressure for social development.

The evaluation of the Fourth Five Year Plan at the end of the 1980s is usually credited with the rethink which led to decentralisation policies. The rise of the central bureaucracy was criticised. It was resented by local elites used to having authority. The king responded by introducing a decentralising policy meant to bring the local public and elites into the development process. The new approach was introduced for the Fifth Plan (1981-86). Each district was to draw up its own Plan, with the District forum (DYT) articulating the needs of the people. Almost 2,000 staff were to be transferred from the centre to the districts to implement these plans. Forty per cent of plan outlays were to be managed by the districts. Moreover, districts were not just bidding for funds; they had to offer to build the schools etc. sought. This was a 'voluntary labour contribution' as distinct from labour taxes (*chunidom*).

The king promoted the decentralisation in tours round the country. Rose (1994) nevertheless argues that 'not much was accomplished in real terms (by decentralisation) in the first few years'. There was resistance from the central administrators, not least to being re-employed in the districts. But there was a further effort to make the decentralised system work for the Seventh Plan (described in Ura, 1994).

I will now give a summary account of basic needs policies.

Education

It is important to remember Bhutan's late start in educational provision and the constraints of a sparse population with few roads. The emphasis has been on expanding primary education, with a goal now on primary education for all by 2000. Measuring success is hampered by inadequate demographic data. High drop-out rates and high repeat year rates remain a big problem. The limited number of places in junior high and high schools is liable to be a source of frustration in the future. Higher education is limited to one college, after which students have to seek scholarships to study abroad. Schooling is free, which adds to the burden of education on the government budget.

Health

Poor health indicators reflect not only inadequate health services but also poor hygiene and use of contaminated water. Policies have included a network of basic health units as well as hospitals in population centres. Manpower is a big constraint to the expansion of services, in spite of use of many Indian doctors and nurses. A notable success has been the achievement of universal child immunisation, which was claimed in 1990.

Agriculture

Most Plan expenditure for agriculture is mainly for small irrigation and land development schemes. In addition, there is an extension service with 233 staff around the country (one for every 280 farming households) and an aim of reaching 400 staff. As with other village-level services getting enough qualified staff is a major constraint. They supply seeds and plant protection. At times the government has been very concerned that the country has to import foodstuffs, but by the Seventh Plan self-reliance was interpreted as being able to sell enough abroad to buy these imports. Similarly, the individual farming household aim is either being able to produce ones own food or having enough income to be able to purchase sufficient food.

Water Supply

From the Fifth Plan, Bhutan has had a big water supply programme. In that Plan period almost 1,000 village water supply schemes were implemented but some were of poor quality and suffered from lack of maintenance. The Sixth Plan involved rehabilitation of 303 schemes and construction of 600 new schemes. However, many were not completed and emphasis was changed from quantity to quality of schemes. The Seventh Plan includes repair of 500 schemes not in working order. The aim is that 45 per cent of the population should have access to safe water from improved schemes by 1997.

Rural Electrification

The Seventh Five Year Plan reports that over 90 per cent of the population of Bhutan do not have access to electricity. There is a programme of rural electrification but progress depends on new generating stations in remote areas as well as distribution. The World Bank (1994) claims a domestic electrification rate of 20 per cent, which it described as 'relatively high'.

'Self-reliance'

Reducing Dependence on Foreign Aid

Since the Fifth Plan, Bhutan has had a declared objective of reducing dependence on foreign aid. The percentage of internal funding for that Plan was 38.1 per cent, compared with 5.4 per cent for the Fourth Plan. After 1988-89 the internal funding for government expenditure has been over 50 per cent (i.e. more than aid receipts). Aid as a percentage of GDP was as high as 59 per cent in 1983-84 but then was reduced to 18.7 per cent in 1989-90 (Seventh Plan data, 1991). A more specific aim is that of full internal funding of the current costs of the government budget. This was achieved in the first two years (1992-94) of the Seventh Plan, but a small deficit is expected thereafter (Report of 5th Round Table meeting, 1995). The argument is that aid should go to investment and not be dissipated in more consumption. It is an argument that can be qualified on grounds of basic needs, or on grounds of the investment nature of consumption expenditure like teachers' salaries.

The planners are finding it impossible to increase domestic revenues at the rate of economic growth and still less at the rate of growth of public expenditure. It is a problem analysed in World Bank, 1994. The key resource project strategy is not producing an elastic revenue flow. The Chhukha Power scheme is a prime example of this, thanks to inflexible tariffs for sales to India. The World Bank argue that Bhutan needs to bring forward further power projects and change its tariffs for power and wood. There was an adjustment in electricity tariffs to India in the mid 1990s which raised revenues.

Taxes represent only a small proportion of government revenue, especially since the Chhukha Project began providing revenues. And of tax revenues direct personal taxes represent only a small proportion. This may well be a significant source of revenue loss. Labour taxes are more significant than income taxes for most households. Ura (1988) estimates that the agricultural sector only contributes 2 per cent of the tax revenue.[8] Unlike some developing countries Bhutan gets little income from international trade taxes, because of the dominance of India, with whom it has free trade, in Bhutan's trade. The Asian Development Bank (1998) report the tax ratio of 8 per cent of GDP is low compared with other countries at similar income levels.[9]

Bhutan has largely succeeded in its policy of diversifying its sources of aid. India's share of aid was 83.3 per cent in 1981-82, 62.1 per cent in 1985-86, 69.4 per cent in 1990-91. The World Bank (1994) figures for grant aid show a further fall in share from India from 60 per cent in 1990-91 to 29 per cent in 1991-92, 38 per cent in 1992-93, and an estimated 35 per cent in 1993-94. These figures exclude non-budgetary grants from India, mainly for defence services, of around Nu 300 million per year. India gives aid in a particularly advantageous form without specifying its use. Of non-Indian aid 35 per cent is from UN agencies, 20 per cent other multilateral aid, 43 per cent bilateral and 1.5 per cent non-governmental.

Although aid is mainly in grant and very soft loan forms, Bhutan does have a growing international debt. Commercial borrowing for industry and sizeable budget deficits in 1988-89—1991-92 contributed to this. The World Bank (1994) reveals that exceptionally in 1994-95 debt servicing is due to be over 40 per cent of current government expenditure.

Bhutan continues to be popular with aid donors. Its documentation about aid requirements (see Royal Govt. of Bhutan, 1991 Seventh Plan, Vol. 2 and Royal Govt. of Bhutan, 1995 Report for Round Table Meetings of donors) is impressive. In 1994, the World Bank was arguing that Bhutan was not reducing its dependence. The Seventh Plan showed more aid being required than for the Sixth Plan. Current expenditure tended to rise faster than domestic revenue and overall expenditure faster than revenues plus grants. But by 1998, the Asian Development Bank was reporting that domestic revenues would cover recurrent costs completely.

Foreign Investment

As part of self-reliance Bhutan has adopted a very cautious attitude to foreign investment. In theory foreign investment is an alternative to aid for funding development. The World Bank (1994) discuss foreign investment as an alternative basis for Bhutan's power sector on the grounds that India may well not be willing to fund further large power stations on the basis they did for Chhukha. One example of foreign investment in a big project is a Japanese corporation's involvement in a large ferro-alloy factory which began production in 1995 (the Bhutan Government keeps a 25 per cent share, according to the Seventh Plan). Nor need the policy just be for big projects. Nepal, for instance, has a lot of Indian inward investment in shops etc. The arguments tend to be different for export oriented developments and those catering for the small Bhutanese market. Bhutan is now permitting private ownership of firms and trying to foster entrepreneurship outside the public sector, so there would seem to be room for further foreign investment.

Employment of Foreigners in Bhutan

Development programmes in Bhutan have made very great use of foreign workers, mainly Indian, both for skilled and unskilled labour. One estimate, from World Bank, 1983, is that 76 per cent of the wage and salary employment in the country was non-national i.e. 35,000 out of 46,000[10]. This includes a higher percentage still of the casual labour, estimated at 28,000, in the country, i.e. those involved in road-building, dam construction etc. Examples of the non-national skilled staffs given in the Seventh plan, 1991, include 1,125 out of 2,625 teachers , 48 out of 85 doctors, 46 out of 202 nurses, 639 out of 1,310 graduates in the Civil Service, and 715 out of 2,576 skilled staff of joint and public sector industries.

The willingness of India to supply contract labour has been a major contribution to development. So has their willingness to provide training places for the Bhutanese. Skill shortages have still been acute, holding up all skill intensive programmes. But since the Sixth Five Year Plan there has been a policy of reducing dependence on foreign workers and accelerating training for Bhutanese. The policy of 'beneficiary self-help' in village public works can also be seen as a response. A national workforce has been created for construction, though productivity is reported to be low and turnover high (Royal Govt. of Bhutan, 1991).

The Seventh Plan has a chapter on Human Resource Development (HRD) and clearly aims to give a high priority to training of Bhutanese. The approach is that of manpower planning via an HRD Master Plan for the country. There are a number of training institutes now in Bhutan, e.g. for teacher training, agriculture and forestry, health science, management and technical training. Of course, to begin with, they have to be staffed by a high proportion of non-nationals. But in addition, the Seventh Plan sought US 50 million dollars for external training and use of external consultants in in-country training. Bhutan was unable to get that amount of aid and has had to cut back the programme.

It is perhaps not surprising that the number of foreigners in Bhutan caused some resentment. This was fuelled by reports that a number of contract workers stayed on as illegal immigrants. Insensitive handling of such then contributed to the rise in tension between Bhutanese of Nepali stock and the remainder at the end of the 1980s.

National Identity : Sustaining Traditional Ways of Life

A number of writers on Bhutan's development praise the way traditional aspects of life co-exist with development. They point to conservation policies in forestry and farming, and the control of tourism (*cf.* Karan, 1987; Samarasinghe, 1990; World Bank, 1989).[11] Bhutan is compared favourably with other parts of the Himalayas, notably Nepal. The relative absence of population pressure is clearly a major factor : hence the addition of population planning policies in Bhutan in recent years.

In the case of forestry, Bhutan is blessed with around 72 per cent (LUP, 1995) of its land area covered by forests. To help preserve this in

1979 the Government nationalised all logging. It has since been developing a programme for Forest Management Units and from 1991 has a Master Plan for Forestry Development. The World Bank 1994 Report generally praises the approach to forest management, though the subsidised availability of wood for house building is criticised for its economic cost.

In the case of agriculture and animal husbandry the World Bank Reports of 1989 and 1994 both stress the delicate ecological balance fostered by traditional systems. This is in contrast to calls to intensify agriculture, such as in the 1983 World Bank Report. One concern is that overgrazing of pastures and forests is occurring because these are collectively owned resources. However, there are traditional means of managing these resources (see Ura, 1993)[12] which it is important that the environmental policy which is being developed takes into account.

Almost invariably development involves people moving from farming to new occupations and from countryside to towns. One of the factors behind 'self-reliance' as a goal (Rose, 1978; Ives & Messerli, 1989)[13] was the emergence of a 'food-deficit' for Bhutan as agricultural outputs with the reduced farming population could not keep pace with demand. But, as we have seen, it was recognised that exports can pay for imports, so that with the development of cash crops Bhutan could allow itself to be no longer self-sufficient in each item of the diet. The cash crops are primarily oranges, apples, potatoes, cardamom, ginger and chillies. Sales of these in India, Bangladesh, Singapore and Japan pay for imports of rice, wheat and edible oils. There are no statistics to tell if farming is seriously losing its workforce, but it would not appear so.

Bhutan has taken a radically different development path than neighbouring countries in the case of tourism. They have foregone a lot of potential income and jobs by deliberately limiting their tourist numbers. This is done by requiring tourists to come in 'packages' at rates controlled by the Government, currently 220 dollars per person per day (with some variations, e.g. for trekking). In addition, tourists are barred from entering most temples and dzongs. The policy looks like one of limiting contact with foreigners — as in the ban on receiving foreign television. Uncontrolled tourism can damage traditional lifestyles, but the Bhutanese approach seems to overcontrol, so that only the rich can visit.

The most radical 'national identity' policies are those of the late 1980s, which are seen by many Nepali citizens of Bhutan as hostile to them. A dress code and language rules have been rigidly applied in much of the country to seem like an 'aggressive policy of cultural assimilation' (Sinha,1994).[14] The Buddhist majority feared being swamped by the Nepalis, but the result has been ethnic strife where there had seemed to be fairly good relations at one time (Rose, 1994).[15]

Conclusions

Bhutan has come a long way in a very short time with its modernisation. Its programmes of rural development are starting to make a real impact on the lives of people of Bhutan (see chapter on education and health). But the country also needs further resource development, if only to pay for the rural development. That also fits 'self-reliance' — reducing dependence on foreign aid and foreign workers. The country rightly sees the need for care in the development of its resources, part of its environmental policies which are an example to the rest of the Himalayan region. It has to be careful too developing its hydro-electric power resources, to see that it gets a proper return.

Above all, Bhutan has to solve its ethnic problems. It cannot afford a large disaffected population in the area of much of the country's economic development.

REFERENCES

1. Ura K., 1994, 'Decentralisation and Development in Medieval and Modern Bhutan' in Aris, M. & Hutt, M., *Bhutan : Aspects of Culture and Development*, Kiscadale.
2. Rose, L., 1978, *The Politics of Bhutan*, Cornell UP.
3. Rose, L., 1994, 'The Role of Monarchy in the Current Ethnic Conflicts in Bhutan' in Hutt M. ed. *Bhutan : Perspectives on Conflict and Dissent*, Kiscadale.
4. Shah, S., 1989, Developing Bhutan's Economy, *Asian Survey*, pp 816-831.
5. World Bank, 1994, *Bhutan : Country Economic Memorandum.*
6. *Kuensel*, April 1995 (Bhutan's National Newspaper).
7. Shah, S., 1989, *Developing Bhutan's Economy.*
8. Ura, K., 1988, *Economic Policy and Development in Bhutan.*
9. Asian Development Bank, 1998, Bhutan in *Asian Development Outlook.*
10. World Bank, 1983, *Bhutan : Development in a Himalayan Kingdom.*
11. Karan, P. P., 1987, *Bhutan : Development amid Environmental and Cultural Preservation*, Institute for the Study of Languages & Cultures of Asia & Africa, Tokyo University of Foreign Affairs.

Samarasinghe, S., 1990, *Bhutanese Economy in Transition, Asian Survey*, pp. 560-575.

World Bank, 1989, *Bhutan Economic Report.*

12. Ura, K, 1993, *The Nomad's Gamble*, National Environmental Commission, Bhutan.
13. Rose, L., 1978, *The Politics of Bhutan*, Cornell UP. and Ives J. and Messerli B., 1989,
    The Himalayan Dilemma, Routledge.
14. Sinha, A.C., 1994, 'Bhutan:Political Culture and National Dilemma' in Hutt, M. (ed.) *Bhutan : Perspectives on Conflict and Dissent*, Kiscadale.
15. Rose, L., 1994, The Role of the Monarchy in the Current Ethnic Conflicts' in Hutt, M. (ed.) *Bhutan: Perspectives on Conflict and Dissent*, Kiscadale.

10

EDUCATION AND HEALTH

— Neil Fraser

Education

IT is important first to emphasise the very late start made in developing mass education in Bhutan. Prior to the third king coming to power in 1952 there were very few schools at all apart from monastic schools for monks and nuns (Rose, 1978)[1]. Other Buddhist countries, e.g. Burma, had mass literacy much earlier. The new king recognised the need for education, in particular to develop the skills for development of the country. By 1959 there were 59 schools with 1,500 pupils. The fastest expansion came after 1981 (Fifth Five Year Plan) when a more determined effcrt to give mass literacy began. The expansion since then has been such that by 1995 there were 288 schools with over 77,000 pupils enrolled (Royal Govt. of Bhutan, 1991 and Ministry of Health and Education, 1995).[2]

The development of education in Bhutan has been very dependent on India, for the funding and the staffing of much of the system. This meant also importing curricula and exams from India. It has also meant a concern that 'rote learning' practices were prevalent in Bhutan. Over the years Bhutan has wanted to reduce this dependency but it is a slow process building up a stock of Bhutanese trained teachers. There has been a large use of untrained Bhutanese staff.

The medium of instruction in Bhutanese schools is English. This means initial years in school are dominated by language learning, especially for those who start their education late (although 6 is the official age for starting school many have been starting much later). English is the commonest subject for repeat years, the amount of which is an acknowledged problem. Using English enables teachers to be recruited in India, but they have often been subject teachers rather than primary class teachers. One unfortunate result has been the introduction of subject teaching from class 4 in Bhutanese primary schools, which Bhutan is now trying to eliminate.

The structure of education consists of Primary Schools, Junior High Schools, High Schools, and a small number of tertiary institutions. These include Sherubtse College which has degree level courses (including lately Honours), 2 colleges for training teachers, and 2 technical colleges. The big emphasis is now on developing primary education with the aim of 100 per cent enrolment by the year 2000. Entry to post-primary education depends on the number of places available, which are not sufficient for the skilled manpower needs of the country, so they need to be expanded too.

The sparse population of Bhutan with a number of small remote villages poses particular problems for education in Bhutan. The initial approach to the problem was known as extended classrooms but this was subject to a critical evaluation in 1990 (Harley and Weber, 1990).[3] Remote communities had to construct these 'extended classrooms' themselves and then the Education Department would supply the teachers, who would be supported by a parent school. In practice teachers were generally untrained and unsupported and these schools offered no progression for pupils beyond the extended classrooms, which usually were only up to grade 3. The report wanted the remote villages to have their own small schools using multigrade teaching to get pupils through all the primary years (as schools in remote places in the developed world do). Bhutan has responded by renaming the extended classrooms community schools. They are to be based on at least 30 students within one hour's walk, with one or two teachers using multigrade teaching (UNICEF, 1992).[4] In 1995 there were 102 community schools with 9,655 pupils (Ministry of Health and Education, 1995). They at most seem to have pupils to grade 4 though Harley and Weber, 1990 suggested to grade 6.

The demand for education in Bhutan is high, as can be seen from the willingness of villages to build their own schools. Nevertheless some children do not go to school and for many others schooling is only a brief episode. It should also be remembered that some 1 in 5 boys go to become monks when aged 6 or 7. Their schooling will be in monastic schools, which are not included in Department of Education Statistics. The aim of the Department is to increase gross enrolment during the Seventh Plan from 67 per cent to 88 per cent (Royal Govt. of Bhutan, 1991), but lack of reliable population figures makes these very rough figures. The main reason given in a survey of parents of non-enrolment are 'needed at home' (64%), 'could not afford expenses' (67%), and 'school too far', (33%) (UNICEF, 1992).[5] Furthermore drop-out rates are high. The value of 2 or 3 years in school must be recognised as not high. Non-attendance too is greater among girls. Parents are less willing to board girls or send them to walk long distances. The Government have set themselves as a target equality between girls and boys — redressing the 1990 ratio of 60 boys to 40 girls in primary schools.

The Government recognises it has a problem of quality in its Primary schools. It is addressing this through the 'New Approach to Primary Education' (NAPE) which involves a new child-centred curriculum no longer dependent on Indian exams. They are wanting to reduce non-national teachers at Primary level but this will mean an expanded output of teachers from the two teacher training institutions, which between them only had 83 primary teacher graduates in 1991 (UNICEF,1992).[6] The World Bank and Swiss Government are funding the expansion of teacher training capacity in Paro. Table 10.1 indicates the very large gap between trained Bhutanese teachers and total teachers at primary level. And as it is a lot of classes in the early years of primary school have 50 or more pupils (UNICEF, 1992).[7]

Table 10.1 : Teachers at Primary Level in Bhutan, 1995

School	Trained nationals	Other nationals	Non-nationals
Community Schools	70	148	10
Primary Schools	280	769	307
Junior High Schools*	67	163	112

* Junior High Schools have more primary level pupils than junior high level pupils, although many teachers will be reserved for the higher grades.

Source : Ministry of Health and Education, 1995.

Plate 15 : Two girls in traditional and modern costumes

Plate 16 : Girls at Bumthang in national costumes

Health

Bhutan has a lot of problems. Indicators like life expectancy, infant mortality, child mortality, maternal mortality, although based on patchy data, are very poor (see Table 10.2). In addition there are indicators of malnutrition, dietary deficiencies and high rates of certain diseases e.g. diarrhoea, respiratory tract infections, worm infestation, malaria, T.B. Finally a high total fertility rate threatens living standards (5.6 births per woman according to Asian Development Bank, 1998).[8]

Table 10.2 : Health Indicators : Bhutan

Life Expectancy	—	—	1990	48.9*
Infant Mortality	1984	142	1994	70.0
Child Mortality (u5)	1984	21.5	1994	9.42
Maternal mortality	1984	0.77	1994	0.37

Source : Royal Govt. of Bhutan, 1991 and 1995.
*World Bank (1994) gives a life expectancy figure of 66 for 1994.

These health problems stem partly from poverty, but they are increased by unhygienic practices. Houses lack facilities for washing people, clothes or food, which tend to be done only irregularly. Stoves, unless modernised, are very smoky. There are many folk beliefs about illness, and traditional healers and monks tend to be consulted before 'western' health services (Wilken & Barth, 1989)[9]. Ninety per cent of births are at home, generally without any trained help. Life-saving operations may be refused for fear of being made disabled.

Health services for Western medicine are a very recent development—mainly from the 1970s. There has been a primary health care emphasis from then — although the hospital system takes up much of the manpower and budget. The main institutions of the primary health care system are 'basic health units' of which there were 72 in 1991. These should each have 3 trained workers but the Seventh Plan (Royal Govt. of Bhutan, 1991) admits that only 40 per cent were fully staffed. A positive account of what they can achieve is given in an article by Morris-Jones (1985)[10]. In addition there are 44 dispensaries and 350 outreach clinics according to the Seventh Plan. But it is important to remember the remoteness of most Bhutanese villages. UNICEF remark that 60 per cent of the basic health units and dispensaries are several hours or days walk from the nearest road (UNICEF,

1991).[11] They also cover only a small number of villages, so to extend further there are some 950 voluntary village health workers who are the first point of contact with the system (though even they are only in 20 per cent of villages). The majority of these voluntary village health workers are male village elders and many are pretty ineffective, according to UNICEF.

The referral units beyond the primary health care system are the hospitals, 16 district hospitals and 3 regional hospitals (one of which, in Thimphu, acts as the national hospital too). The 85 doctors in the country, of whom 48 are foreign, work in and from these hospitals (figures for 1991 from Royal Govt. of Bhutan, 1991, but the Statistical Yearbook for 1990 gives 157 doctors in 1989). Referrals to hospitals are hampered by poor communications in the country, plus the delay often in going to basic health units etc.

The health system that has been developed does have some successes. A major one is the immunisation programme, organised particularly through the outreach clinics. By 1991 the programme claimed 100 per cent success (based on more than 85 per cent uptake). The progress in reducing high levels of infant and child mortality can probably largely be attributed to this programme. A study of primary health care and infant mortality in a poor area of East Bhutan over the period 1984 to 1991 shows what was achieved (Bohler,1994).[12]

A lot of outside attention to Bhutanese health has been directed at mothers and children, as in the work of UNICEF(e.g. Levitt and Doma, 1993).[13] The high levels of reported infant, child and maternal mortality are good reasons for this attention. Fortunately there is not the discrimination against girls found in much of South Asia (this lack of discrimination includes the Nepali Hindus of the south of Bhutan, according to Wikan,1991).[14] It seems that the primary health care system can claim some success in services for women and children and that women have been prepared to attend clinics. More education of girls will also help health, as will better diets which should come with higher living standards. UNICEF have urged the development of a midwife service and the aim declared in the Seventh Plan is to expand the proportion of deliveries attended by trained people from 12 per cent in 1990 to 50 per cent in 1997.

Up to now there have been a number of particular programmes trying to deal with particular problems in Bhutan, e.g. for diarrhoea, TB, malaria and also population planning. The Seventh Plan declares that these programmes are now to be incorporated in what should be an integrated primary health care system.

Health spending as a proportion of the national budget is reported by UNICEF (1992) as declining from 8 per cent in 1970 to 3.8 per cent in 1990 but then being increased back to 8 per cent during the Seventh Plan. As noted in the Development chapter government expenditure in Bhutan is now tightly constrained although basic needs programmes tend to be favoured within the constraint.

Conclusion

Bhutan's rapid development of education and health services is now facing the problems of such development. Reduced mortality rates are producing a high population growth rate, making for overcrowded schools and pressure for jobs from school leavers. Social service expenditure is said to be rising alarmingly with the cost of the ambitious targets for access to schools and health services in a country of sparse population, very poor communications and small numbers of skilled people to deliver these services. The country now has a taste for these basic needs of rural development.

One of the things threatening these developments is the troubles in the south of the country. Many facilities there were closed in 1991. Schools and health services need to function for the future of the area (the area of much of the wealth of the country). It is also the area where malaria is prevalent and in an increasingly virulent form.

REFERENCES

1. Rose, L., 1978, *The Politics of Bhutan*, Cornell UP.
2. Planning Commission, Royal Govt. of Bhutan, 1991, *Seventh Five Year Plan, 1992-97* and Ministry of Health and Education, 1995, Royal Govt. of Bhutan, Education Division, *General Statistics*.
3. Harley, B. and Weber C.M., 1990, Extended Primary Classrooms in Bhutan, Report submitted to the Department of Education, Royal Govt. of Bhutan, and UNICEF.
4. UNICEF, 1992, *Children and Women in Bhutan, 1991, A Situation Analysis*.
5. *Ibid.*
6. *Ibid.*
7. *Ibid.*
8. Asian Development Bank 1998, *'Bhutan' in Asian Development Outlook*.
9. Wikan and Barth, 1989, Bhutan Report, Results of a fact-finding mission,

UNICEF.

10. Morris-Jones, 1985, 'Tales of the Unexpected : The Basic Health Unit in Bhutan', *British Medical Journal,* 291, 28th September 1985, pp. 877-878.
11. Wikan, 1991, *The Girl Child in Bhutan,* UNICEF/University of Oslo.
12. Bohler, 1994, 'Has Primary Health Care Reduced Infant Mortality in East Bhutan ? The Effects of Primary Health Care and Birth Spacing on Infant and Child Mortality Patterns in East Bhutan', *Journal of Tropical Pediatrics,* 40(5), pp. 250-260.
13. Levitt, M. & Doma, U. 1993, *Motherhood in Bhutan,* Redd Barna and UNICEF.
14. Wikan, 1991, *The Girl Child in Bhutan,* UNICEF.

11

ARCHITECTURE, ART, RELIGION AND LANGUAGE

— *Anima Bhattacharya*

Architecture

THE world of architecture, art, craft, music and language are still distinctive in Bhutan. Bhutan's architectural forms are quite diverse. Temples, monasteries, dzongs (fortresses), chortens, stone walls, palaces and mansions, and dwelling houses make up a unique architectural landscape.

Dzongs, the Bhutanese fortresses, were constructed at strategic points for administrative safeguard. They contain both monastic communities and administrative offices of the district governments. Architecture of Dzongs at times is claimed as originally Bhutanese, though there are differing opinions as to the influence of Tibetan architecture in these cases. Dzongs are imposing structures with several stories usually not exceeding six, with spacious halls, impressive alcoves, passages, corridors, wooden panels, stairs, galleries and paintings. Many of them are situated beside streams while others occupy a commanding high position in the locality.

Temples (Lakhangs) are fairly small buildings of simple design which are likely to comprise one storey around a small enclosed courtyard. These buildings seem to have been the first forms of religious architecture and some of them are centuries old. They differ from ordinary houses by the red band painted on the upper part of their walls and an ornament of gilded copper on the roof. Inside, the walls are completely covered with paintings and the interior space is sometimes divided by pillars into an antechamber and sanctuary proper. Temples are maintained by a caretaker who may be a member of the owner's family if it is private, or be assigned by the State Clergy if it is State property.

Monasteries (Gompas) are mainly of two types, namely cluster type and dzong type. The 'cluster' type believed to be older consists of a core formed by one or two temples and dwelling structures are grouped around. Dzongdrakha in the Paro valley, Phajoding in the Thimphu valley, and Tharpaling in the Bumthang valley are examples of this type.

The solidity and elegance of sloping walls in the Dzongs and settled areas are noteworthy. The detailed woodwork and the ethereal character of the pitched roof make the dzong among the most beautiful architecture of Asia.

Lordly mansions seem to have appeared during the period at the end of the 19th century when the country began to experience relative peace and the lords of Bumthang acquired great political power. Their layout was, in fact, very similar to that of a fortress. The lord and his family lived in a central building surrounded by an enclosed courtyard with service buildings backed up against its walls. However, the architecture of these residences was less severe than that of dzongs, which are built for defence. The upper floor of the central building was always turned into a private chapel. This room was decorated with painted murals and contained numerous statues as well as religious books needed for rituals. The most noted examples of this type of architecture are in the Bumthang valley, south of Tongsa, and in the Paro valley.

Chortens representing Buddha's mind, are receptacles of worship, equivalent to an Indian stupa where relics of Buddhist divines were deposited, erected in memory of an eminent Lama or to ward off evil spirits from places normally considered dangerous, such as crossroads, bridges and mountain passes. There are three types of Bhutanese chorten : large chortens of whitewashed stone are modelled after the chortens of Bodhnath in Nepal; smaller chortens very much in the Tibetan style are found especially in Central and Eastern Bhutan and often protected by a wooden superstructure. Their main architectural characteristics are that they have a square base on which the main structure rises with diminishing breadth. They are closed on all sides and contain the sacred relics deposited there. The Paro chortens, the third type, are pagoda-like in basic style; they have neither doors nor windows. The square-shaped chortens with the roof composed of four slopes and the upper part just below the roof decorated by a wide red stripe are mostly found in Western Bhutan.

Chortens are consecrated and contain statues, religious books, fragrant herbs and even weapons. Two chortens may be linked together by a stone wall which are relatively few in number and fairly short in Bhutan, a circumstance perhaps explained by the topography.

Bhutanese religious works describe eight kinds of chortens *viz.* (i) a prayer-shrine; (ii) a centre of great enlightenment; (iii) of good fortune; (iv) of peace and tolerance; (v) of Buddha's return to earth from heaven; (vi) of dissemination of religion; (vii) of harmony and victory; and (viii) of Mahaparinirvana of Buddha. Their architectural style varies, but some contain frescoes of ursurpassable beauty representing teachings of Buddhism.

All dzongs, lordly residences and important temples are built of stone, while village houses are constructed of different materials depending on their locations. In western Bhutan, cob (rammed earth) or wood is the commonest building material, whereas in the centre and east, stone or wood is the choice. In Eastern Bhutan, woven bamboo matting is also used for building, and often serve as roofing for small houses on stilts.

Throughout Bhutan houses have the same chracteristics : they are rectangular with one or two storeys. Windows traditionally had no permanent protective screening; sometimes bamboo screens were put up to shut out bad weather without excluding light, but today glass is used in the more populated areas.

The roof is the same as for other buildings, and the space between roof and the two-sided sloping roof is used for drying vegetables, meat and for storing hay. In towns, this space is closed off with bamboo mats.

Art

Almost all Bhutanese art is symbolic, a rare blending of Tibetan, Indian and Chinese traditional styles is characteristic Bhutanese setting. The effect of real Indian influence is somewhat less marked than that of Burma and Siam, which entered by way of Assam and was, therefore, often taken to be Indian influence.

Plate 17 : Residential houses-cum-shops in Bumthang town

Plate 18 : Palace of the legendary King Sindhuraj at Bumthang, now residence of a lama family

Religious themes predominate all Bhutanese art forms; liturgical and iconographic limitations, however, do not make it static. It is rich in colour and detail, and dynamic in its exclusive representations of Bhutanese religious concepts, belief and way of life.

The art and industry of Bhutan cover a wide range of activity, beginning with artistic production of objects of gold, silver, copper and including production of steel weapons like swords and daggers. Bhutan is well known for its articles of brass and bronze as well. In Varanasi, gold and silver smiths have excelled in producing designs in white and yellow by intertwining gold and silver in picturesque patterns. This work of great art is known as Ganga-Jamuna art of India. The same technique but with Bhutanese setting and pattern has been flourishing in Thimphu, and other centres of culture like Paro from ancient times.

Another industry of Bhutan centres round the manufacture of casting bells of very fine tones. The composition used for best bells includes a large quantity of silver. The largest bell is not more then 24 inches in diameter .

The Bhutanese women have an old tradition of producing not only woven strips of cotton, but also sheep's wool which is rain and wind proof. Weaving is a national industry of every household. In fact, every home has a room devolved to weaving in which wool, silk and cotton textiles are produced along with floor carpets.

Beautiful needle-work pictures of saints produced on hanging banners are another speciality of Bhutan. It appears innumerable pieces of coloured silk and brocades are applied in a most artistic manner with elaborate stitches of all kinds. This is, indeed, a veritable work of art.

Another artistic industry is based on the use of split cane. Apart from basket weaving, there is also the production of mats, finely woven of the same material, namely, split cane. They are reported to be delightfully fine and soft, so flexible that they can be rolled up into quite a small space and they are durable.

Mention must be made of the importance of wood carving in Bhutan. According to tradition, sandalwood statues from India were specially sent to some of the shrines in Bhutan to be consecrated as objects of worship. Wood sculptures are usually made of pine and walnut. Wood-work, wooden bowls and receptacles — mainly from the Tashi Yangtse region in Eastern Bhutan — are turned on foot-powered treadle lathes. Bowls lined with silver are used for butter-tea or alcohol. Wooden receptacles are used as serving dishes for food and also as plates by the heads of the household.

Some bowls and all receptacles are lacquered black or red with a substance that is extracted from the tree *Rhus succedanea*. This is the same substance that the Japanese use for lacquer, but in Bhutan the finishes are less sophisticated.

Frescoes and Murals

The genius of Bhutanese art is best represented by its frescoes and murals. Its primary aim appears to be the visual enunciation of the Buddhist *tantric* tradition, mythology and popular interpretation of religious philosophy, beliefs and concepts. Religious motifs, therefore, dominate most of the Bhutanese frescoes and murals. Guru Padmasambhava with his manifold incarnations is the idea of all Bhutanese painters. The mural showing him in his various emanations in the Tagtshang monastery is one of the finest specimens of Bhutanese paintings. The Bodhisattava frescoes adorn the walls of the Chari monastery. The Paro Lhakhang contains some of the finest frescoes of the early Tantric Buddhist tradition, and also now virtually extinct Tibetan form of Bonism or Shamanism.

The frescoes in the Sangak Chhoekar Monastery, Paro, however, are purely Buddhist in form and concept representing scenes from Buddha's past incarnation from the Jatakas. Some frescoes depict peaceful as well as wrathful Tantric deities as they appear in intermediate state between death and rebirth. One of the murals in the Simtokha Dzong shows the guardians of the Deva Loka in their awesome majesty.

The Thankas

The thankas are painted scrolls — cotton cloth, canvas or silk brocade applique, or embroidered paintings of religious themes. Bhutan's Thankas are usually large, and depict an artistic skill and a rare exquisite finesse. Some are painted, while others represent applique work on silk brocade. The great Thanka at the Paro Dzong done in silk brocade applique contains the central figure of Guru Padmasambhava and his various incarnations. The Pha-Jo-Ding Monastery in Thimphu possesses one of the largest and exquisitely painted Thanka ; another large scroll hangs on the top storey walls of the National Museum in Paro.

The Mandalas

The *mandala* or mystic circle murals represent Bhutanese Buddhist concept of the cosmogony of the universe. The frescoes of the gigantic *mandala* to the left of the eastern gate in Paro Dzong is a classic enunciation of this metaphysical concept of the universe. The mystic spiral is encircled by a red flame. Other circles in blue, red, green, yellow, black and white go spinning around, revolving and intersecting each other. These represent generally the Buddhist conception of the manifold worlds and astral systems in the universe.

Mandalas of various mystic patterns abound in monasteries and Lhakhangs throughout Bhutan. They are an external part of both religious initiation and meditation. They are painted on Thankas, on temple walls, on ceilings of national buildings and represent the best Bhutanese artistic skill in rich colour and design executed by artists. By far the best specimens of the *mandala* art are depicted — one on the ceiling of the Tshogdu Hall, the other on a Thanka of the five-fold of geometrical designs of exquisite beauty and colour with Shabdrung Rimpoche on the left and Kunkhyen Pema Karpo on the right.

The Wheel of Existence

By far the entire concept of births and rebirths — the most fundamental laws of Buddhism — is illustrated in frescoes of the pictorial 'cycle of existence'(*Bhavachakra*) in the temples. In the Wangdiphodrang Dzong the *Bhavachakra* or the wheel of existence is the favourite theme of painting. Three great sins are represented by hog (ignorance), cock (passion), and snake (hatred). Salvation is inconceivable without the complete annihilation of these three sins. The pictorial wheel of existence illustrates how to obtain redemption from the ceaseless, endless cycle of births and rebirths. Three cardinal sins — lust, anger and stupidity — are the causes of rebirths (*Nidanas*).

These frescoes also illustrate the six forms of rebirth depicted in the inner circle of the 'wheel of life', *viz.* (i) in the *Devaloka* or the world of gods; (ii) the *Asuraloka* or the world of the demons ; (iii) the *Tirjyakloka* or the world of beasts; (iv) the *Pretaloka* or the world of ghosts; (v) the *Narakaloka* or the hell, and (vi) the *Manusyaloka* or the world of mankind.

Plate 19 : Three Chartens on the way to Paro : Bhutan, Tibetan and Nepal Types

Plate 20 : One of the earliest monasteries with flying Prayer Flags at the entrance : Bumthang

Dance

Bhutanese classical dance is reflected by their religious mask-pageants and ritual dances. Traditionally, most of these were initiated first by Terton Padmalingpa in the 15th century and then by Shabdrung Rimpoche in the mid-17th century, as accompaniment to prayers to the protector god Mahakala in the Punakha Dzong, which is famous as reincarnation of Jampel Dorji started a pattern for prayers to the protector goddess Shri Devi in the Thimphu Dzong. During the prayers, the lamas have to dance for destruction of enemies of Buddhism and for bringing peace to the country and its people. Among the celebrated religious dances are : (i) the Shanag Chham or the black hat dance; (ii) the Degyed Chham or the spirit dance; (iii) the Shinje Chham or the Yamaraja dance; (iv) the Le-goen Chham or the dance of the Protector God; (v) the Lhamo Tsokhor Chham or dance of the Protector Goddess ; (vi) the Serdeng Ber-kor or Procession dance; (vii) the Goenpo Mang-Chham or dance of the Protector God and others; (viii) the Dur-Dag Chham or dance of the Shamashan Lord; (ix) the Tum Ngam Chham or the angry mood dance; and (x) Guru Tshengyed or the dance of the Eight Gurus.

Most of the religious dances are symbolic and have a common theme : to destroy and trample underfoot the evil spirits. The swords of the dancers cut through ignorance, their drums drive away all malevolent evils and demons. The stag-dances represent the conversion of evil spirits taking a vow to protect *dharma*, the irate death-masks remind the humanity that death is inevitable to all mortals, and that the life is illusory, that the *nirvana* can only be attained by good deeds.

Bhutan's semi-religious and lay ritual dances are all said to have been organised by Guru Rimpoche, Ningmapa Lama, Tsangpa Jare, and Shabdrung Rimpoche. These ase performed by non-monks on annual festival days even in remote villages all over Bhutan. Bhutanese people living undisturbed for more than a thousand years in their mountainous valleys still possess folkloristic treasures having their roots in the dawn of history. This was possible because Bhutan, often called the last 'Shangri-la', was a closed mountain paradise, and indeed a closed country. The art of folk dances, as old as Bhutan, is a special art of the Bhutanese practised in every remote village, or in courts of the dzongs, under the guidance of a dance master who acts also as a medieval choreographer during festivals. These layman dancers introduce the religious festivals performed either on the New Year which, according to the moon calendar, is February, or on the foundation anniversaries of a monastery. On these occasions the religious mask dances as well as other dramatic plays are shown which date back to the end of the 8th century. The costumes and the precious wooden masks belong to the monasteries. They are stored in the temple of the Protective Deities and given out only for dance ceremonies. The accompanying music for the religious dances is played by an orchestra of monks having big and small drums, long and short trumpets and clashing cymbals. The layman dances are accompanied by old songs which also give the rhythm to everyday work. Songs of Central Asia Gesar Epic are still alive in Bhutan.

The interest in and liking for music is showing noticeable inclination towards Indian and Western music. Time will see how traditional music is preserved and exotic music incorporated.

Religion

The original 'Bon' populations were sprinkled all over the country prior to the migration of the Tibetans in the seventh century and later of the Tibeto-Burmans from the east. These 'Bon' people developed a religion based on superstition and demonolatry and were administered by the local chieftains.

Religious concept and practices changed from Bonism to different shades of Buddhism from the eighth century onwards. Bonism virtually belonged to Animism based on nature worship. This Bonism had its rituals which were mainly concerned with the eradication of the demons and other evil spirits who were considered to be responsible for misery and sickness of mankind. Some aspects of this nature worship were incorporated in the Mahayana Buddhism introduced by Guru Padmasambhava. Thus, early Buddhism was essentially imbued with strains of hereditary tantricism. Gradually this Mahayana Buddhism evolved into refined Buddhism, incorporating Brahminical sciences. Consequently many Hindu gods and goddesses were deified in the concepts and principles of metaphysics and religious philosophy of Buddhism in Bhutan. It was, in fact, through Tibet that Indian Brahminical tantricism may be said to have left an indelible mark on Buddhism practised in Bhutan. This evolutionary process must have been a continuous one through the establishment of various religious sects in the country.

The earliest Buddhist sect to come to Bhutan from Tibet was the Nyingmapa sect. The Nyingmapa sect was followed by the adherents of the Sakyapa sect. However, from the eleventh century onwards, the Kargyupa sect had started migrating to Bhutan. They were the venerated adherents of Saint Naropa Mahaguru. A branch of the Kargyupa school, named the Drukpa sect, finally established the State religion in Bhutan. The historic scholar Pamkarpo (1527-92) was a gifted writer and systematised the teachings of the Drukpa. The Drukpa sect of Buddhism today forms the real foundation of the national church of Bhutan. It is estimated that there are about 5,000 lamas in the country. They dominate the religious life of the people, devote their entire life to prayer and meditation, and depend upon the charity of people. The centre of activity of the lamas is the monastery, and almost every village of Bhutan has a monastery of its own. This set-up of religious organisation has not changed much in recent times. The only remarkable change that has taken place is observed in the decentralisation of the responsibility of imparting education to the people. The monasteries and the dzongs were the only seats for education in the past. So these institutions were the centres of education, studies and libraries for the local and regional population. This state of affairs has changed with the introduction of modern education on Indian and Western models during the last few decades.

Language

The chief language of Bhutan is Dzongkha, which is taught in all schools and written in classical Tibetan block letters Ucăn for the purposes of print. Although Dzongkha may have its origin in classical Tibetan, it varies greatly in pronunciation. Prefixes and suffixes, which became mute in high Tibetan languages, are still spoken or at least heard in Bhutanese pronunciation. In the southeast of the country, the Mon and Khen languages of the aboriginal inhabitants of the Himalayas are still alive. These old languages are a striking linguistic mirror of the rich Bhutanese tradition comprising treasures of local legends, songs, plays, and are a lively echo in which still resound the voices of sunken pre-historic times.

Numerous dialects that are spoken in Bhutan give one a confused impression. Dialects exist all over the world and are subsidiary to the main spoken languages. As far as Bhutan is concerned, there are four known main languages as follows :

(a) Dzongkha — spoken only in western and northern Bhutan, is also the official language of the country;
(b) Bumthangkha — spoken in central Bhutan;
(c) Sarchapkha — spoken in eastern Bhutan; and
(d) Nepali — spoken in southern Bhutan.

For the first three languages, the script used is Tibetan Ucăn. The Nepali language is written in Nagri. In urban areas Hindi is not unknown. English is extensively used in the offices.

Six dialects of the country belong to the Dzongkha group. They are as follows :

(i) Dzongkha — the language of the dzongs, especially spoken in west Bhutan and the common language of the schools;
(ii) Tapadamteng — in the southwest near Buxa;
(iii) Laya Lingzhi — the shepherd's language in the northwest in Laya and Lingzhi;
(iv) Tshalingpa — in the Tshali area and in east Bhutan;
(v) Dagpakha — a Tibetan dialect of northeastern border;
(vi) Mira Sagtengpa — Tibetan dialect of Tashigang in the farthest east of Bhutan.

Seven dialects belong to the Khen languages and are found in the following parts of the country :

(i) Mangdikha — in the Mangdi or Tongsa district;
(ii) Gungdekha — an archaic language of the south;
(iii) Khenkha — in the Khen region south of Tongsa;
(iv) Bumthangkha — especially in Bumthang and in the whole of central Bhutan;
(v) Tsamangkha — in the east-north-east on the eastern border of Kurto;
(vi) Kurtopakha — especially in Kurto; and
(vii) Salabekha or Yangtsepakha — in the southeast Tibet close to Bhutan border.

All these dialects have no scripts of their own, but they are now transcribed into Ucăn. Furthermore, there are two dialects belonging to the Mon group. These are :

(i) Monpakha — in the south and southeast; and
(ii) Sharchagpakha — in the east and southeast, especially spoken in Tashigang and Dungsam.

Namthars or Biographical Literature

A vast mass of historical and biographical literature grew up in Bhutan during the 17th and 18th centuries. They may be described as historical works of great literary merit. They record the pre-historic traditions and historical legends of Bhutan-Lho-Mon before the introduction of Buddhism into Bhutan and its relations in the 7th century with Tibet by Srongtesen Gampo, Thrisong Deotsen and Thiralpachen. They also record the coming into Bhutan of the foremost disciple of Kargyupa sect, who propagated and spread the tenets of the Kargyupa sect in Lho-Mon kingdom before the advent of Shubdrung Ngawang Namgyal into Bhutan.

Soon after the death of Shabdrung Ngawang Namgyal, the literary trend in Bhutan became predominantly biographical and the geshes began to write namthars. This biographical literature of Bhutan is considerable in its output. The literature of Bhutan consists of numerous other biographies which are still extent.

Bhutan's language and dialect are rich in vocabularies which is manifest in the archaic dialect of herders of yak in Ha and sheri cattle in Bumthang. The west and central nomads have vast repository of terms for cattle and calves of differeat colours, age, sex and breed and the like. This naturally speaks of great richness of their dialect. It is indeed not a matter of easy affair to refer to a cattle or calf in one word describing its colour, age, sex and breed.

12

THE PEOPLE AND THE SETTLEMENTS

— *Anima Bhattacharya*

The People

Ethnicity

THE average Bhutanese people are healthy, roly-poly, sturdy or hardy — predominantly of Mongoloid strain. As there is paucity of information regarding the ancient indigenous and immigrant population of Bhutan, perhaps no authentic account can ascertain the early population complex of Bhutan.

The Mongoloid stock consists predominantly of Indo-Mongols, Tibeto-Mongols and Tibeto-Burmans. Bhutan's relation with India is from great antiquity. In those days of shifting boundary and lack of strong monarchical or theocratic rule, Bhutan was often treated as a part of Cooch Behar or Assam in India in the pre-historic period. Later the country was divided into small territories ruled by local chiefs. These territorial chiefs were engaged in constant strife with each other till the establishment of monarchy in the seventeenth century.

Tibeto-Mongols are handsome, tall with stout or robust physique. Indo-Mongols or Tibeto-Burmans are not as robust as Tibeto-Mongols but possess basic mongoloid features and healthy disposition.

The nomenclature of the country and history of many places are suggestive of close link with India like Tibet and besides Indo-Mongol population, Indo-Aryan strain also may be observed in the population mix.

The coming of Buddhism from Tibet to Bhutan in the seventh century introduced a marked change in the population complex of the country. Tibetan King Srongtesen Gampo stretched his influence from west to central Bhutan as far as Bumthang by establishing two Buddhist temples, one at Paro and another at Bumthang. The temples still exist in dilapidated condition and regular prayers and festivals take place there.

Prior to the inroads made by Tibetans, the country was inhabited by numerous tribes. Chief among them were Sha of Wangdiphodrang valley and Wang inhabiting Thimphu valley. Other tribes were Shashebs of Tongsa and the Togas of Dungna. Lepchas inhabited western Bhutan.

Minor tribes like Brokpas, Dakpas and the Sherchokpas were spread over different parts of Bhutan. Nepalese consisting of the Rais, the Gurungs, the Limbus, the Bahuns, the Chettris and the Tharus now dominate southern Bhutan.

On lower rungs of the social order, the Goens, the Kajibs and the Thowzibs — mostly herdsmen — inhabited the Punakha and Thimphu valleys, where also lived the Pewa community selling betel leaf (*pan*) and spirit, and the Zongzibs who were menials. The Bhutanese with Koch blood were known as Thop. Besides these, the Dahyas in Chamurchi, the Bagboras and the Oles in Chirang and the Totos in Luckepur are mentionable.

People of Indo-Aryan origin flowed into southern Bhutan from Bengal and Assam, forming a small minority with distinct language, religion and culture. The Indo-Mongoloid people came to Bhutan from the east and settled in the southeastern part of the kingdom. They also have a distinctive language, religious practice and way of life.

The coming of the Tibetans or the Tibeto-Mongols is coeval with the coming of Buddhism to this country in the seventh century. Mongoloid immigration of Tibetan origin was reinforced by the Milongs group in the ninth century and also by the influx of Buddhist Lamas. Milongs established their settlements in the valley of Lho. Buddhism got a setback during the reign of the apostate king Lang Darma who reinstated Bonism and prohibited the practice of Buddhism. Lang Darma's inglorious anti-Buddhist activities lasted only for a short time — the incidence led to large scale migration of lamas and the Tibeto-Mongols.

Buddhism declined with temporary dominance of the Bon priestly class. Corruption and superstition infiltrated the religion until its revival by Atisha, Gerapsel and Renchhen Zangpo in the eleventh century in Kham and central Bhutan. Buddhism regained its supremacy in Tibet and the Himalayan region giving birth to numerous schismatic sects and schools of thought. The earliest unreformed sect was Nyingmapa followed by the reformed Kadampas, the Sakyapas and others.

Gelukpas or the Yellow Hat sect was founded by Tsongthapa (1359-1419). This sect drew its spiritual and doctrinal inspiration from Atisha (1042). Atisha introduced strict moral code for the monks, and gave a central organisation to Buddhism in the Himalayan region. He advocated celebacy and moral abstinence, and deprecated the practice of magic arts. He set a reformist movement which culminated in the establishment of the Kadampa sect and a few centuries later Tsongkhapa made it more ritualistic and less ascetic. It became the dominant Buddhist sect in Tibet.

At present, the social order is composed of three distinct classes of people, namely the ruling class, the priestly class and the farming class with assisting people of various services.

Population

Bhutan is a land of scanty population. The idea of population is clouded by confusing figures as there was no regular institution of census in the recent past. The first census of Bhutan was made in 1969. The figures derived from various sources state that the population vary from 1 to 1.1 million. Authentic sources claim that the actual figure is 6 lakhs (600,000), which is much less than the previous figure.

Nature has divided the people of Bhutan into three cultural zones — namely southern, central and northern Bhutan with distinct agricultural practices and products, biodiversity and language mix. The population primarily concentrates in the central part laterally with elevation between 1,524 m (5,000 ft) to 2,438 m (8,000 ft) and above. Similarly, there is another zone of concentration in the foothill region bordering India with same lateral pattern from west to east. In the north, above 2,743 m (9,000 ft), population is sparse forming the third lateral zone from west to east.

In north-south direction demographic Bhutan can be clumsily divided into three zones, like Western, Central and Eastern Bhutan. Western Bhutan clearly lies west of Sankosh river. Central and Eastern Bhutan remain overlapping between the rivers Sankosh and Manas in the east. East of Sankosh river begins Central Bhutan but its eastern boundary is indeterminate. In all probability, Eastern Bhutan lies east of the tributary of Manas named Kuru Chu, another landmark being Mongar dzong. River valleys here are the main forces of north-south division rather than altitude affecting population distribution.

The distribution of population in 18 districts or dzongkhags is not even at all which is revealed in the following Table 12.1.

Table 12.1 : Distribution of population in Bhutan by districts

District or Dzongkhag	1980	1985	1990
1. Thimphu	64,300	71,674	80,684
2. Chhukha	97,200	108,347	121,964
3. Ha	17,100	19,061	21,457
4. Paro	39,800	44,364	49,941
5. Samchi	138,400	154,272	173,666
6. Chirang	104,500	116,484	131,128
7. Daga	27,700	30,877	34,758
8. Punakha	34,500	38,457	43,291
9. Wangdiphodrang	47,700	53,170	59,854
10. Bumthang	23,600	26,307	29,614
11. Sarbhang	112,800	125,736	141,543
12. Shemgang	43,300	48,266	54,333
13. Tongsa	28,600	31,880	35,888
14. Lhuntshi	36,900	41,132	46,302
15. Mongar	71,300	79,477	89,468
16. Pemagatsel	35,100	39,125	44,044
17. Samdrupjongkhar	72,200	80,480	90,597
18. Tashigang	170,000	189,496	213,318
Bhutan	1,165,000	1,298,605	1,461,853

Source :Statistical Year Book of Bhutan 1990 : Population projections based on 1980 census data.

In every district of Bhutan, population appears to rise considerably from 1980 to 1990. Population is the largest in the spacious district of Tashigang and is minimum in Ha, small in size and have a high elevation. Tashigang is followed by Samchi, Sarbhang, Chirang and Chhukha in aggregate population of noticeable size. The previous Table stating total population as 1,461,853 in 1990 is an estimated population projection based on 1980 data.

The following Table 12.2 gives an outline of population by age and sex for 1990.

Table 12.2 : Population by Age and Sex in Bhutan, 1990

Age-Group	Males	%	Females	%	Persons	%
0-4	115,901	7. 93	115,216	7.88	231,117	15.81
5-9	95,543	6.53	95,870	6.56	191,413	13.09
10-14	71,570	4.90	71,083	4.86	142,653	9.76
15-19	75,104	5.14	72,872	4.98	147,976	10.12
20-24	68,616	4.69	66,148	4.52	134,764	9.22
25-29	58,931	4.03	56,713	3.88	115,644	7.91
30-34	50,137	4.43	49,198	3.37	99,335	6.80
35-39	42,751	2.92	41,211	2.82	83,962	5.73
40-44	36,104	2.47	34,361	2.35	70,465	4.82
45-49	30,409	2.08	28,354	1.94	58,763	4.02
50-54	26,504	1.81	24,900	1.70	51,404	3.52
55-59	22,136	1.51	20,979	1.44	43,115	2.95
60-64	17,255	1.80	16,371	1.12	33,626	2.30
65-69	12,440	0.85	12,119	0.83	24,559	1.68
70-74	8,026	0.55	8,053	0.55	16,079	1.10
75 and above	8,406	0.58	8,572	0.59	16,978	1.16
All Ages	739,833		722,020		1,461,853	

Source : Statistical Year Book of Bhutan 1990 : Projected from 1980 population census.

In almost all age groups, males outweigh females population excepting in the age groups 5-9, 70-74 and 75+ . This phenomenon is suggestive of higher death incidence among male children (5-9) and women survive better than men in the highest age groups. Population steadily decreasing in higher age groups indicates lower expectancy of life.

Though the male population outnumbers the females in both the periods (1980 and 1990), the figure is expected to improve considerably during the later period. Similarly, population will slightly improve in children and working age group indicating viability. On the other hand, aged population has declined in the decade indicating loss of maturity and wisdom in the society.

Table 12.3 : Population Indicators, Bhutan

Population Indicators	1980	1990
Sex-ratio (Males/Females)	104.3	102.5
Age Composition %		
0-14	38.5	38.7
15-64	57.2	57.4
65 and over	4.3	3.9
Dependency ratio		
Child dependency	67.3	67.4
Old Age dependency	7.5	6.8
Index of Aging	11.2	10.1

Note : Index of aging refers to the number of persons aged 65 and over as perce..tage of the number of persons aged 0 to 14.

Source : Statistical Year Book of Bhutan 1990.

Dependency ratio is quite large on the human resources in the working age group as child dependency, old age dependency and old age index together presses the mentioned age group hard.

Table 12.4 : Measures of Fertility and Mortality in Bhutan

Fertility	
Crude birth rate	39.1
General fertility rate	169.6
Total fertility rate	5.9
Gross Reproduction rate	2.9
Net Reproduction rate	1·9
Rate of Natural Increase (%)	2.0
Mortality	
Crude death rate	19.0
Infant mortality rate	142.0
Life Expectancy at birth :	
Males	45.8
Females	49.1

Source : Statistical Year Book of Bhutan 1990.

The analysis of the data mentioned in Table 12.4 is given as follows:

Crude birth rate : number of live births in a year per thousand population;
General fertility rate : number of live births in a year per thousand women aged 15 to 49;
Total fertility rate : average number of children born by women during their lifetime;
Gross reproduction rate : average number of daughters born by women during their lifetime;
Net reproduction rate : reproduction rate adjusted for mortality of women;
Crude death rate : number of deaths in a year per thousand population;
Infant mortality rate : number of deaths during first year of life per thousand live births;
Rate of natural increase : the difference between the crude birth rate and the crude death rate.

Marital Status

Marital status of the population of Bhutan varies naturally with age groups. In the status of "never married", the age group of 0-14 constitutes 99.9 per cent (males) and 99.8 per cent (females). This status decreases with progressing age groups in which the number of males exceeds the number of females.

In the "married" status, the number of women is much greater than the number of men in the lower age groups i.e. 15-19, 20-24 and 25-29. Above these age groups the males outnumber the females upto 50 and over age groups.

The incidence of divorced or separated spouses is not negligible which is higher in cases of women in upper age groups, beginning from 25-29 to 45-49, reaching peak values at the ages 30-34 (7.3%).

Among the widowed population, widows again outnumber the widowers from the age group of 25-29. Bereavement of spouses is maximum in the age group of 50 and over. The figures for the widower in the age group is 15.7 per cent and widows 32.2 per cent.

Table 12.5 : Marital Status and Age of Population in Bhutan (%), 1984

Age Group	Never Married		Married		Divorced Separated		Widowed	
	M	F	M	F	M	F	M	F
0-14	99.9	99.8	—	—	—	—	—	—
15-19	92.1	73.4	7.7	25.5	—	—	—	—
20-24	54.5	30.1	43.8	64.8	1.5	4.4	—	—
25-29	26.8	12.4	70.0	80.5	2.4	5.7	—	1.4
30-34	13.1	8.4	82.8	81.3	1.9	7.3	2.3	3.1
35-39	10.0	5.5	85.8	84.9	1.8	5.7	2.5	3.9
40-44	7.6	6.3	87.9	81.4	1.7	5.3	2.8	7.0
45-49	6.9	6.9	86.8	78.7	1.4	5.4	5.0	9.1
50 and over	6.5	4.2	75.4	58.8	2.3	4.8	15.7	32.2
All Ages	57.8	52.0	38.0	39.0	1.0	2.8	3.2	6.2

Source : Statistical Year Book of Bhutan 1990 : Demographic Sample Survey 1984.

The Table 12.5 for marital status is trite and commonplace for the never-married and married population and also for the widowed population that can be observed in most Asian countries. Widowed population is again indicative of lower expectancy of life for the males. The anomalous male-female data for the divorced/separated status points to the practice of polygamy in the social fabric.

Manpower

The nature of Bhutan's manpower requirement has been gaining sophistication and complexity. Further, it has become imperative that we take stock of the national manpower planning encompassing even that of the private and the agricultural sector.

Bhutan is in the process of seeking the UNDP/ILO assistance in this regard, to help in formulating a new manpower policy and to develop an appropriate manpower planning system.

The past manpower plans have continuously been giving priority to education and training at both degree levels for studying engineering including architecture, medicine including traditional Budhhist medicines, agriculture, animal husbandry and forestry, teaching, nursing etc. These are to meet the manpower requirement to fill up the operational posts and duties in the agencies.

Subjects such as environmental studies are of recent origin but its importance is gaining urgency and popularity. Under the current HRD programme, fellowships for social forestry, forest inventory, pest management, forest management, forest fire protection, entomology, wild life census, watershed management and land use planning were awarded. Of these only the watershed management course leads to a degree, the rest being short term courses. These will be guided in projecting the requirement of the secretarial and the concerned sectors in incorporating the national and zonal plans in the next plan period.

It is estimated that the future demand shall focus not only on the field requirement where such knowledge is applied but in other areas as well such as research and development, policy and direction, and teaching of the subjects in Bhutan's context.

Our world as described by the World Commision on Environment and Development is as follows :

> "A small and fragile ball dominated by human activity and edifice but by a pattern of clouds, oceans, greenery and soils. Humanity's inability to fit its doings into the pattern of changing planetary systems, fundamentally. Many such changes are accompanied by life threatening hazards. This new reality, from which there is no escape, must be recognised and managed."

Prerequisites for Higher Studies

To pursue a course of environmental studies one should prepare oneself for the study of science, geography and mathematics at the ICSE, ISc and degree levels.

Career Prospects

Career in the category of policy and direction and teaching in environment opportunities other than the few posts envisaged for creation in Bhutan in the near future are as zoologists, wildlife botanists, wildlife ecologists, entomologists, nature conservationists, animal behaviourists and lecturers/trainers etc.

Human Resources

In Bhutan there are few statistics available on human resources and labour. The only data available are some broad level data on economic activity which was obtained from two household surveys, and some data on civil service employment were obtained from administrative records.

The economy of Bhutan is characterised by the predominance of people engaged in self-employment, particularly those working their own land. There is also a good deal of small cottage industry often undertaken in the home. The number of persons working for wages is relatively small.

In 1984, over 93 per cent of the population aged 15 to 64 were engaged in some economic activity, including 90 per cent of the males and over 96 per cent of the females. For males, 78 per cent were engaged in agriculture and 12 per cent in other activities, while for females, over 95 per cent were engaged in agriculture and only 1 per cent in other activities.

As for July 1990, the total civil service employment amounted to 12,000. Of these, 10 per cent were females. There has been a substantial reduction in the number of civil service employees since January, 1987 as a result of the Government's policy for a small, compact, efficient government. Table 12.6 gives an idea about the occupational structure in Bhutan.

Table 12.6 : Occupational Structure in Bhutan (%), 1984

Age Group	Agriculture	Govt. Service	Business	Others	No Occupation	Total
Less than 15	8.2	—	—	0.6	91.3	100.0
15-19	72.4	1.2	—	1.9	24.4	100.0
20-24	83.9	5.8	1.3	2.0	7.0	100.0
25-29	88.0	6.8	1.1	2.2	1.8	100.0
30-34	90.8	5.4	1.1	2.1	0.6	100.0
35-39	91.7	3.9	1.4	2.5	0.5	100.0
40-44	93.5	3.0	1.4	1.9	—	100.0
45-54	94.4	1.7	1.0	1.9	1.1	100.0
55-64	89.6	0.6	0.6	2.0	7.2	100.0
65 and over	72.1	1.0	—	0.5	26.0	100.0
All Ages	57.2	2.0	0.6	1.4	38.8	100.0

Source : Statistical Year Book of Bhutan, 1990.

The Settlements

In 1990, the total number of Gewogs (block made up of several villages) in Bhutan were 191. They were of the following order in the 18 Dzongkhags or districts (Table 12.7).

Table 12.7 : Distribution of Gewogs (Blocks) in Bhutan

	Dzongkhag or District	Number of Gewog or Block
1.	Thimphu	7/10
2.	Chhukha	11
3.	Ha	5
4.	Paro	10
5.	Samchi	16
6.	Chirang	12
7.	Daga	11
8.	Punakha	14
9.	Wangdiphodrang	15
10.	Bumthang	4
11.	Sarbhang	15
12.	Shemgang	7
13.	Tongsa	5
14.	Lhuntshi	8
15.	Mongar	11
16.	Pemagatsel	7
17.	Samdrupjongkhar	11
18.	Tashigang	20

Source : Statistical Year Book of Bhutan, 1990.

Bhutan is the most scantily populated country in South Asia. Most of the population is concentrated in the valleys and large areas in the north of the country are virtually empty except for nomadic herders.

Bhutanese still live in villages in an extended family system or maintain a strong link with their rural families. The average size of the household or family is estimated to be 8.5. The number of houses per village varies from 2 to 100 with an average of 43. Thimphu in western Bhutan is the capital with an estimated population of 40,000-50,000. The other main urban settlements are Phuntsholing, Paro, Geylegphug and Samdrupjongkhar. All these centres are border towns, excepting Paro, between India and Bhutan. Towns are developing in all the 18 district headquarters.

Rural Settlements

In general, rural houses are highly scattered. The dispersal of houses has many advantages for the households in terms of proximity to arable and grazing land, and forest areas for the collection of firewood. Settlements are more clustered in the colder alpine regions than in the sub-tropical parts of the country. The houses on average tend to be bigger in the central and western Bhutan than in other parts of the country due to climatic and historical reasons as well as availability of building materials. The rural houses are built with strong emphasis on traditional architecture. The average size ranges from about 56 sq m (600 sq ft) to about 84 sq m (900 sq ft) and are mostly two storeys high. The upper storey is used for living and the lower storey is used in most cases for storage of equipments and for keeping domestic animals.

Rural Housing Design

Rural houses are spacious, beautiful and structurally sound using building materials that are locally available. Like other elements of Bhutanese culture, the continuity in the architecture gives the communities a sense of identity. However, rural houses have many deficiencies with regard to health, hygiene, sanitation and convenience. There is inadequate insulation so that space heating is not efficient. Because the chimneys are not well designed, smoke is a major source of respiratory disease. There are no suitable bathing and waste disposal facilities.

Rural Water Supply

Gravity-flow water supply schemes of small and medium size (not exceeding 10 km in length) will continue to be the (low cost) choice of technology. In new schemes only sources with less than 10 faecal coliform/100 ml will be used. The water quality of the existing schemes will wherever needed be improved. Priority will be given to the provision of water supplies to the social institutions. Wherever possible these will be established as part of a community water scheme. In all stages the department will aim at generating responsible community ownership of the completed facilities. The main strategy will be to assist the communities to develop the necessary technical and organisational skills to maintain and operate the facilities once they are created. The involvement of the women will be emphasised. Since mothers and women have set the habits of the young in health and hygiene, the sanitation programmes will focus strongly on the women. The village voluntary health workers or other designated persons will be made responsible for testing the quality of water.

A number of rural water supply schemes have been constructed with inferior designs. Over the years, the schemes have deteriorated and about 500 water supply schemes are not in working condition. These schemes will be repaired with improved designs. There will be periodic monitoring to ensure proper maintenance of the schemes.

Both the Department of Works and Housing and the Department of Health Services will promote the construction of household latrines. Beneficiaries will be free to choose from three different types of latrines (ventilated improved double pit latrine, pour flush latrine, long drop composite latrine) as alternative to traditional latrines.

The improved stoves programme is designed to popularise the appropriate types of stoves for different parts of the country, especially where firewood is deficient. Such improved stoves will save cooking time and fuel and contribute to the control of respiratory diseases.

The water and sanitation construction programme will be accompanied by education campaigns to raise awareness of the importance of this programme in improving health. Communication materials like films, slides, booklets will be produced to reinforce public consciousness about sanitation. The Department will continue to collaborate in the Information, Education and Communication for health programmes.

The Royal Government of Bhutan has constructed 18 model houses in 18 districts to demonstrate the benefits of new designs, building materials and construction methods. Carpenters and masons from rural areas have been trained in order to spread the new skills. The main thrust of the new initiative was not on government investment in construction of new rural houses but on redirecting the rural population's efforts to construct functionally efficient, cost effective, comfortable, hygienic and durable houses.

Urban Settlements

Urban settlements are classified as having a population between 500 and 25,000. In 1989, 32 urban centres were classified as towns.

Urban scene in Bhutan deserves special attention with their sparse distribution and small population clusters compared to the urban concept in India, Southeast Asia and other parts of the world. The percentage of urban population as indicated by the official data was 13 of the total population in 1980 which did not lag much behind the great neighbours like West Bengal (26.49%) and India (23.71%). The distribution of urban centres easily confirms the strategic location of the national capital, the district headquarters in the central part of Bhutan, securely located beyond either Tibetan borderline in the north and Indian border in the south. Such locations are further accentuated by river confluences where river valleys present favourable space consisting of flat surfaces for promising agricultural activities and urban development. Such centres are Ha, Paro, Thimphu, Punakha (old capital). Wangdiphodrang, Tongsa, Jakar in Bumthang, Mongar and Tashigang from west to east. South of this string of urban centres are two rows of towns, one along the border of India, and another in the north of these border towns. From west to east the border towns are Sibsoo, Samchi, Phuntsholing, Sarbhang, Geylegphug, Samdrupjongkhar and Daifam. Between the border towns and the central urban string are Sambey Dzong, Daga, Damphu, Shemgang, Pemagatsel and Deothang. In the northern part, there are only five towns, namely Gasa, Laya, Lunana, Luntshi and Tashi Yangtse. Of these five urban centres, Laya and Lunana are ensconsed in the high solitudes of the Great Himalayas.

Urban expansion in the form of population input and urban sprawl has mostly occurred in the national capital and other interior sections, but border towns like Phuntsholing , Geylegphug and Samdrupjongkhar have swelled noticeably with trade activities as distributing centres of various commodities. Among these, Phuntsholing has expanded amazingly along the river Torsa within the last two decades almost from its obscure tiny existence. Moreover, these towns serve as commuting centres for passengers, tourists , service and technical personnel. New towns have come up with administrative and commercial needs and old towns with expanded national activities have also grown in their bulk. Consequent changes in urban amenities are also taking place with timely necessities and opportunities.

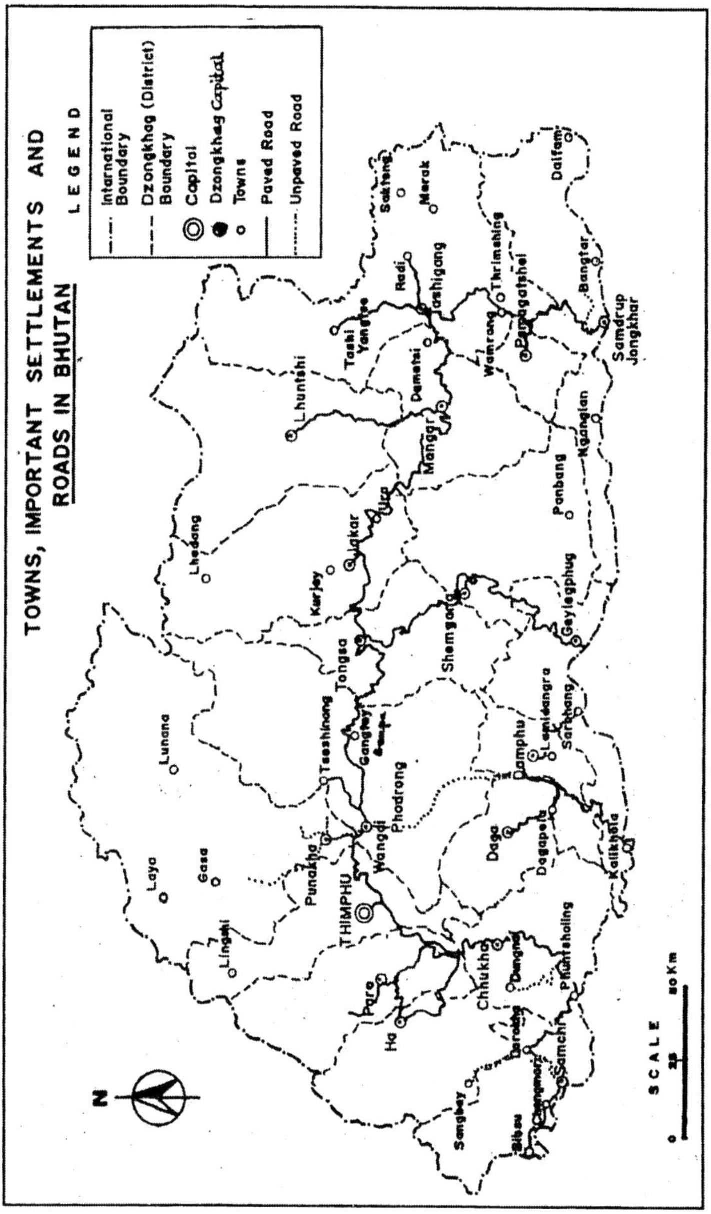

Fig. 12.1 : Towns, Important Settlements and Roads in Bhutan

and opportunities.

The traditional urbanscape of Bhutan was distinguished by spacious living, with distinctive housing pattern with nailless wooden and bamboo structures and the wattle-wall painted with indigenous soil paints. Such wattle-wall houses look strangely alike brick-built houses and the old sentiments of Bhutanese people for bamboo mats can still be traced in the preservation of mat-panel below the sloping roofs.

Recent changes are marked in congested housing provisions and the architecture and the building designs and materials, changing to flatroofs, multistoreyed buildings of exotic types, hollow bricks structures and the use of metals and chemical paints. Certain modern residential private and public buildings look awkwardly bizarre amidst the graceful old houses of traditional Bhutan.

Urbanisation of Bhutan has not been a smooth affair for its rugged terrain, paucity of agricultural land, inhospitable climates and scanty human resources. The rugged hilly terrain has equally rugged experience in the attainment of urbanhood. This mountain country had developed the nucleus of clustered village settlements which turned into urban nuclei in course of time in every dzongkhag or district around the massive dzongs which served as the regional religious, administrative, defence and cultural centres. There is evidence of such urban development with the establishment of dzongs in the seventeenth century, the earliest of which is Simtokha dzong built in 1629. Other dzongs came into existence in quick succession like Punakha dzong (1637), Tashicho dzong (1641), Paro dzong (1646) and so on, giving rise to towns and townlets fitting the timescale of history. At the beginning they were all overgrown villages with agrarian economy which gradually changed to conventional urban economy with administrative, commercial and other cultural activities. At present there are as many as 32 urban settlements of which Thimphu and Phuntsholing are the largest centres.

The towns of Bhutan are mostly located on river basins, where the rivers make their sustaining life arteries. There are instances of their extension and development on hill slopes, but generally, towns have developed on one side of the river and where bridges facilitate communication, towns extend to some extent on the other side. The shapes of towns vary according to the alignment of building spaces of the town's environment and in this respect, each town shows its individuality in its layout.

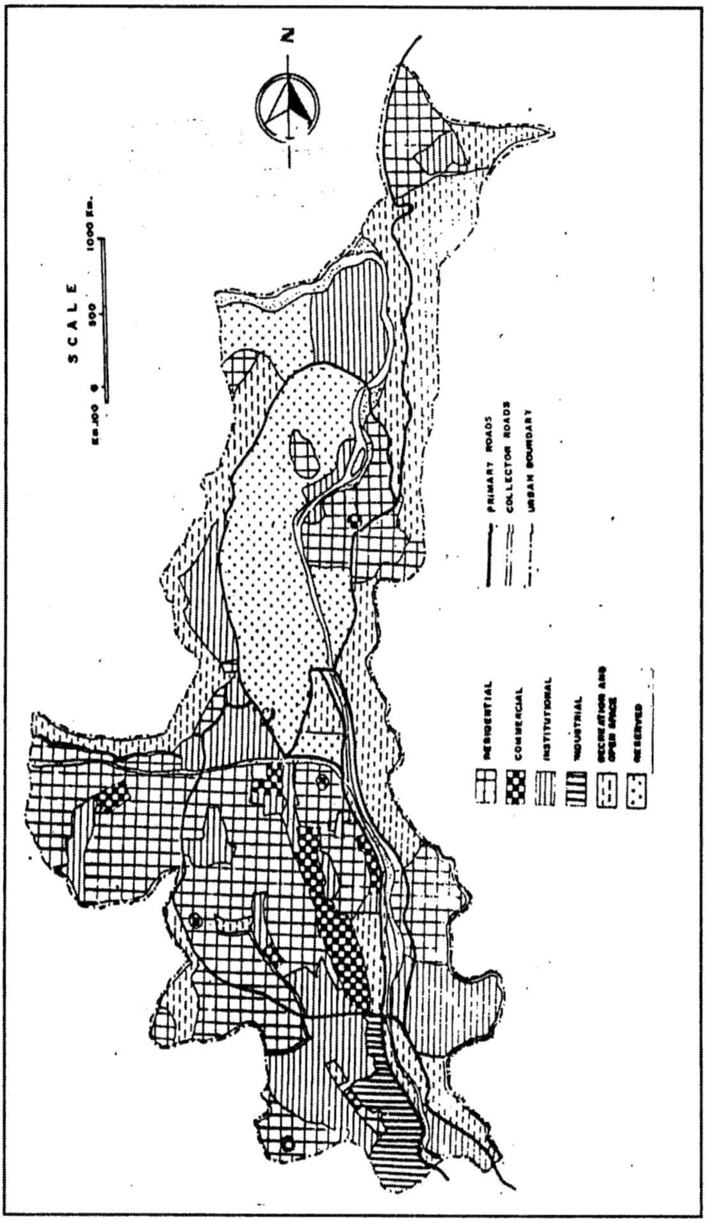

Fig. 12.2 : Thimphu : Major Landuse and Roads

It is necessary to say something about the life and landscape in the two larger towns of Thimphu and Phuntsholing.

Thimphu, the capital city, is situated in the west central Bhutan in the narrow valley of Thimphu Chu flowing in north-south direction. The Thimphu valley stretches for about 11 km from Dechencholing in the north to Simtokha in the south. The valley has an average width of 1 km in east-west direction. The existing urban area constitutes 830 hectares in the central portion of the valley, concentrated in the south of the Chubachu river.

The town spreads over the river flats and also on the adjoining highlands, not too steep for habitation. Later expansion has taken place on the lower plains of the river and also on the highlands in the vicinity of the township.

The old nucleus of the city consists of Tashicho dzong with its administrative enclave, Royal palace and residences of the royal entourages and the trading resident-communities along the Narzim Lam (Main Road). The Royal family, the nobility, the religious sects and the trading families in the past formed the early urban communities of Thimphu.

Thimphu has been enlarged since 1960 with more administrative, educational, trading, constructional and cultural activities in the dzong area, the commercial centre and around the royal residence. From the city centre to Simtokha dzong, the city is building in a depressed manner.

The city is surrounded by agricultural and forest lands which restrict its natural expansion by prohibiting encroachment on the valuable productive land. Other limiting factors are the rivers and steep slopes.

On the whole, Thimphu is a beautiful, quiet urban settlement at an altitude of about 2,440 m (8,000 ft.), originally laid in a spacious manner with sufficiently wide streets and a narrow entry point from the south across a wooden bridge. South of the bridge, open roads lead to Paro, Punakha, Wangdiphodrang, Tongsa and further east and north. Open spaces including vacant lands punctuate the entire city in a dispersed manner which constitute 32.5% of the city area. It is desirable that these spaces are not encroached upon for building purposes. These along with surrounding forests and agricultural land form the formidable environmental shield of the city. If the aesthetics of these open spaces are enhanced, Thimphu will wear a more beautiful look in future.

The buildings in general were two storeyed and occasionally three storeyed with pleasant traditional designs, closely built along the commercial hub. Late structures consisting of residential and institutional types have a variety of designs with exotic, traditional, and a combination of both (Fig. 12.3 to 12.6). National designs naturally have a high aesthetic and utilitarian value with sloping roof, nailless wooden structures, bamboo matting, wattle plasters and the like. The mixture of modern architecture has at many places spoilt the original beauty to a great extent, though they might have improved the internal amenities.

Being a capital city, Thimphu has a sizeable residential area within its bounds. The residential area constitute 38.3% of the total, forming the largest land area among other uses. The administrative (9.3%), commercial (4.0%), health and education or institutional (10.1%) area of the total land uses are significant for well thought-out town planning. Similarly, security (3.8%) and industry (2.0%) do equally have their own reasonable shares in the distribution of the total land area.

The total land area under transportation is not available in the data dossier, but as has been mentioned earlier, the streets are sufficiently wide for a hill town with a well-built terminal at the entry point and parking spaces in the commercial core and elsewhere. With the expansion of the city around the palace area, it will require more bridges across the river Thimphu chu.

By all standards, Thimphu is not a crowded and noisy city, with generally a clean environment where life rolls on peacefully without usual city tensions, anxieties and agitations. Though Bhutan is not visibly an affluent country, Thimphu city does not show beggars, street dwellers, and utterly destitute persons. Nor does the city witnesses frequent scenes of disabled and diseased persons as is found in Indian and Asian cities. People, in general, have a good physique and health. They usually are not lazy and have disciplined working habits. But to lead a really clean life, Thimphu needs more water in every house and institutions and also hot water supply with the development of electricity.

General traffic poses no serious problem till now in Thimphu. The thoroughfares are sufficiently wide for a hill town. But with the rise of vehicular traffic and car ownership, Thimphu will need betterment of the existing roads, pavements, road signs to match international standard. For internal movement of city buses for schools, dzong officials and officials for other institutions, the transportation facilities are satisfactory. But for the movement of heavy traffic constituted by buses and trucks for the growing city, a ring road will be necessary in near future.

Fig. 12.3 : A traditional shop of mainly wooden structure along Norzim Lam, Thimphu

Fig. 12.4 : A Residential Concrete Building with traditional architecture

Fig. 12.5 : A Residential Concrete Building etc.

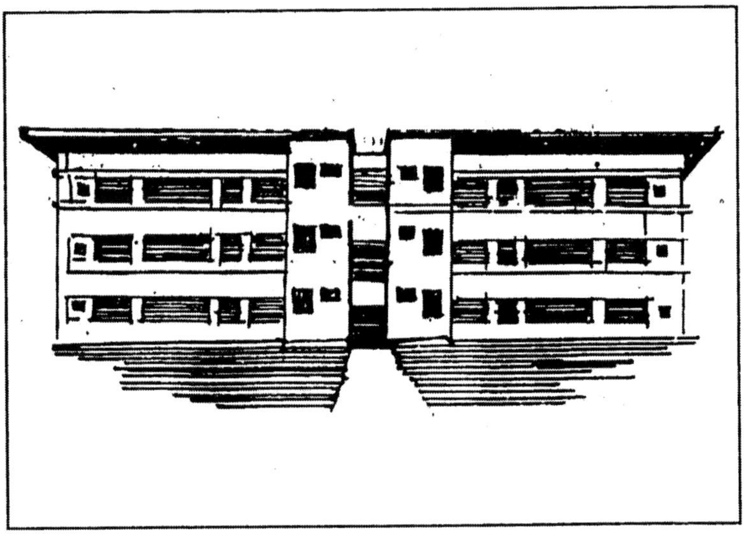

Fig. 12.6 : A Residence of Combined Architecture etc.

For pedestrian traffic, many by-roads, connector roads and narrow roads must be improved; and also the pavements for the pedestrians should be built properly.

Moreover, for growing vehicular traffic, the city will need more bridges across the rivers including one in the neighbourhood of the bus terminal in addition to the existing one.

Phuntsholing

Twentyfive years ago, Phuntsholing was a tiny, picturesque urban settlement in the southern fringe of Bhutan in the alluvial fan zone of the west of Torsa river. By the last two-and-a-half decades, the town has grown phenomenally into the largest commercial entrepot activities for interior Bhutan, India and beyond through Jaigaon, the adjoining Indian commercial settlement separated at one point by a narrow brick drain not even a metre across between the two towns lying in two sovereign states.

It was an ecstasy for an onlooker of the plains for the first time to have a glimpse of Phuntsholing from the entry point of Bhutan gate. Small is beautiful and that was a pleasant truth about Phuntsholing. It is still beautiful though has lost much of its beauty for some crowdedness. The town has sprawled in three directions, north, west and east. In the south, the Indian border puts a limit for southward expansion to the west and southwest, the Torsa is a natural barrier for expansion in that direction. In the north and northeast, rise in altitude restricts the township to grow that way and also to conserve the forest resources to check landslips and soil erosion.

In spite of rapid growth, Phuntsholing has developed quite systematically with proportionate distribution of land area for different uses. The town contains 320 hectares of municipal area to the southeast slope of Kharbandi checkpost, including 180 hectares of corporation area in the west and southwest. As one moves from Bhutan gate to the north and east, the physical landscape is pleasantly undulating culminating into steep slopes futher north (Fig. 12.9).

Phuntsholing may be divided into four sectors of development. Sector I consisting of the intensely built central business district offers little scope for new development, although opportunities for redevelopment exists. For instance, the western half of this sector may still be used for more intense development. Sector II comprises residential zone lying to the east of the

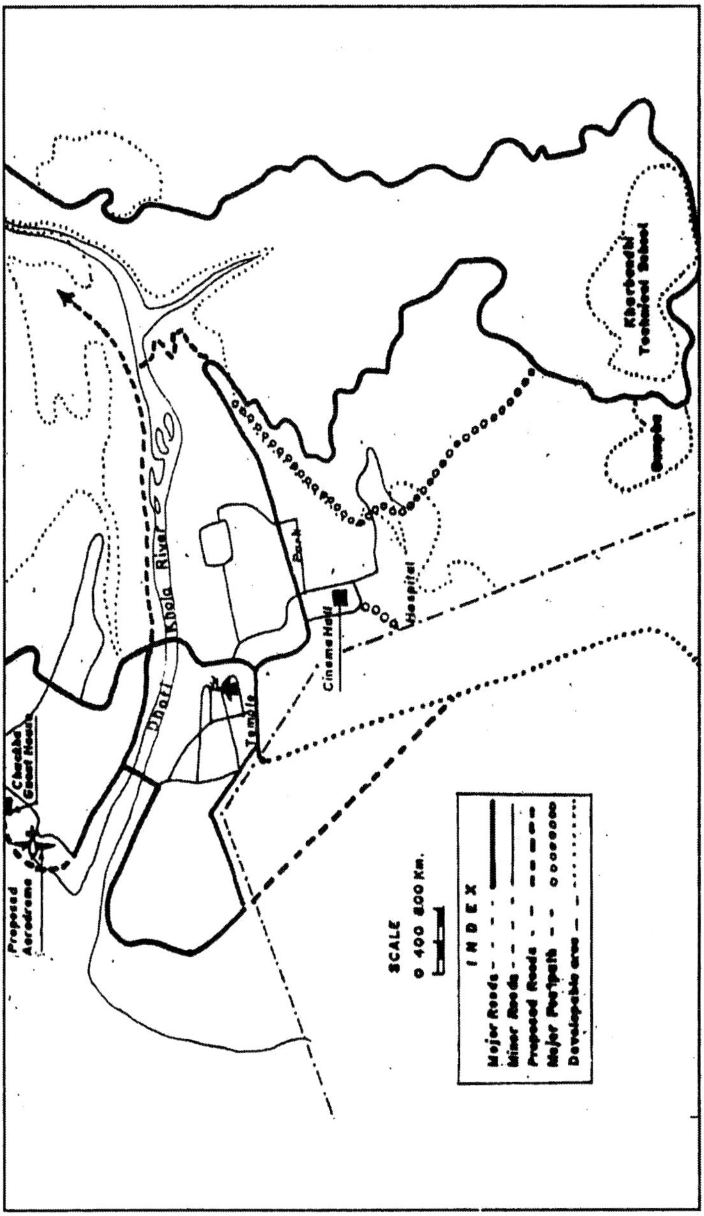

Fig. 12.7 : Phuntsholing : Transportation and accessibility to important places

commercial area and south of the Dhoti Khola river. This sector has scope for consolidation and shift of existing uses for new development. To the north of the Dhoti Khola river lies Sector III, used mainly for industrial purposes. There is ample scope in this sector for residential development and commercial uses. North of the industrial estate, the land is suitable for low rise houses for the unstable nature of the landform.

Sector IV is situated in the south of the city and has small pockets of development. It has a low density of population with substantial area developable for low density institutional purpose of Kharbandi. In 1986, the land use pattern of Phuntsholing was characterised by larger share of residential use (23.5%) followed by roads and rivers (19.6%), institutional (18.8%), industrial (12.6%) and commercial (7.0%) uses. Besides these, special areas comprise 16.8% of the total area. Within the town area, open spaces amount to 17%.

Housing survey in 1986 shows a total of 3,357 households that occupied 2,675 housing units. Residential units include 117 huts and 164 residential-cum-commercial units. The overall average size of a housing unit within the city is estimated at 35.3 sq m (380 sq ft). Small units in both public and private sectors, up to 18.6 sq m (200 sq ft) constitute about 61.6 per cent of the total. Larger units exceeding 92.9 sq m (1,000 sq ft) constitute 11.4 per cent in the private housing units as against 3.6 per cent in the public housing stock.

House ownership rate is low in the town. Only 9.4 per cent of the urban households own their residences. The quality of urban houses is not very satisfactory when judged by the amount of living space. About 25.5 per cent households are shared accommodation. National housing policy, however, prefers separate dwelling which cannot be maintained now for domestic and financial reasons. Household amenities survey in 1986 shows that electricity, piped water, bathrooms, latrine, septic tank are provided in most of the houses, from 73.0 to 86.3 per cent. Piped water and electricity are provided to more than 80 per cent, i.e., 86.3 per cent and 83.0 per cent respectively. The provision of septic tanks is the lowest (73.0%), which is a serious concern for public health.

Urban facilities available in Phuntsholing for the residents are satisfactory for offices, shopping centres, banks and recreation, but for schools, hospitals, bus stands, hotels and restaurants, accessibility is less convenient. But Phuntsholing being small in area, no facility falls beyond 1 km distance.

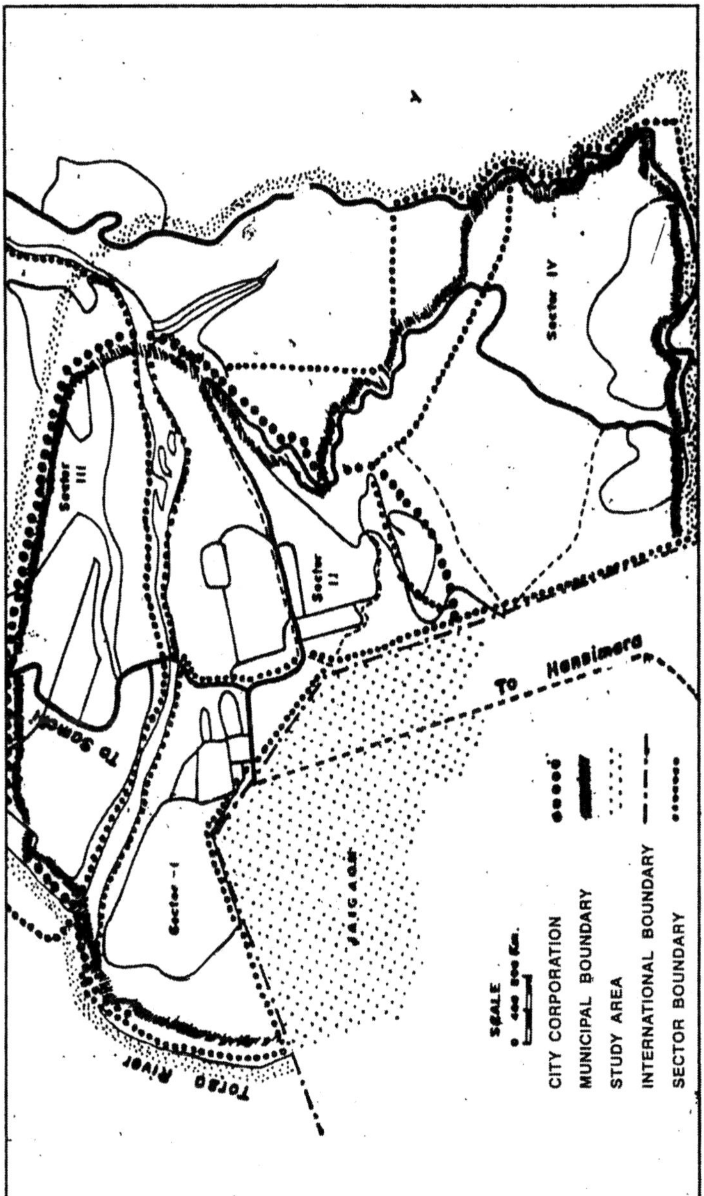

Fig. 12.8 : Phuntsholing —Internal Division

The average household density of the town is 4.5, and the room density varies between 3 and 6 and above. This explains the degree of crowdedness which is quite uncomfortable for a decent standard.

Compared with Thimphu, Phuntsholing has some problems with regard to transportation and traffic. Phuntsholing is essentially a pedestrian town and vehicular traffic is problem-ridden only on the main road at the entry point. Being a border town and serving as an entrepot, the town has sizeable heavy traffic involving long distance bus services and trucks with slow and fast moving vehicles. The narrow point of entry at Bhutan Gate (Fig. 12.9) from Indian border and southern Bhutan and Kharbandi checkpost to interior Bhutan, have much traffic congestion with waiting buses and trucks seeking entry permit and custom clearance. Some alternative arrangements are necessary at this entry point particularly at the Bhutan Gate.

Fig. 12.9 : Bhutan Gate — Phuntsholing

The means of transport in Phuntsholing consist of bus, cars, scooters, bicycles, carts etc. The mixture of slow moving traffic like hand drawn, horse and bullock drawn carts including bicycles has its own character in this pedestrian town which creates congestion at Bhutan Gate. With the rise of car ownership in future, the main roads and the connector roads will require some planning for parking places and movement of traffic. The town requires good pavements along all the roads for major pedestrian traffic.

For the smooth-running of traffic in this growing commercial centre, certain organisation and development are necessary in future. One more entry point besides the Bhutan Gate is to be created for diverting present congestion. To introduce air services in this town having international connections, the proposed airport by the bank of river Torsa, north of Dhoti Khola should be given due consideration. An alternative east-west highway to the existing national highway should be constructed north of Dhoti Khola by-passing the central area which will meet the national highway in the east. Passenger and cargo traffic should preferably be segregated in future. With further development of the town in north direction, some bridges will be necessary for easing traffic movement. With all these parameters of organisation and development, Phuntsholing will have better infrastructure for future development.

Water supply and sanitation conditions are yet to be properly developed. Water for the corporation area is derived from Dhoti Khola river and for Kharbandi area from a tributary of the Dhoti Khola by gravity feed system and distributed untreated. The Kharbandi storage tank serves the school, the hotels, the gompa, the general hospital and the local residential areas which appears to be adequate. But the existing capacity of 3.5 m litres a day for the corporation area should further be enhanced. The wastage and transportation of water across the border to India should be stopped.

Phuntsholing does not have any sewerage system. Individual buildings are served either by a septic tank or by a soak pit. Fourteen per cent of the households do not have any water connection and 27 per cent households are without septic tanks. Lack of water and particularly treated water are the major sources of health hazards. Apartment buildings usually have one common septic tank. Soak pits are common in low density residential and institutional areas. Effluent from the septic tanks flows into the storm drainage system in the densely populated central area which is another concern for the town's health and sanitation.

Garbage collection and disposal is operated by a fleet of trucks, tractors, low-loading covered trailors, closed containers, and wheel barrows. Refuse is collected from the dwellings and the streets to a central collection point to be removed for disposal on the bank of river Torsa. Measures should be taken to check river water pollution in future.

Phuntsholing is a deviation from the rest of the prominent urban centres of Bhutan. It grew up in the border as a carefree settlement without any shield of a dzong. However, a small gompa with a royal guest house beyond the town is the symbol of cultural landmark of the settlement, upholding the religious grip on the social life of Phuntsholing. The original nucleus of the township was a well laid out form performing simple trade and passenger traffic for a non-ambitious nation. The present volume of passenger and cargo traffic of international character is a result of Bhutan's entry into the United Nations and the entrepot activity of a busy commercial centre. Future expansion of Phuntsholing will depend upon the country's relationship with other countries and the necessity of having a larger city at the border of a friendly neighbour. At the present level of requirement, Phuntsholing requires stabilisation without worrying about future expansion.

Changes in the urban sector of Bhutan is aimed at planned development under the aegis of the Five Year Plans. Among these water supply system, internal road network, street lighting, pool housing system, low cost housing, solid waste disposal, sewerage scheme and sanitation feature prominently in the programme of urban development.

It appears from whatever official data furnishing facts about the country's march towards urbanisation, that Bhutan has set her foot on the right track for future urban development and whatever mistakes have taken place in the beginning, will, in course of time, be eradicated if the Plans sincerely take shape.

The author is tempted to add some facts about a few more urban settlements derived from her visual experience through personal and field visits. The author's first visit in the inner Himalaya of Bhutan was in the eastern part of the country in 1979 through Samdrupjongkhar to Kanglung and Tashigang *via* Rangya in India.

Samdrupjongkhar

A border town in the eastern Bhutan for entry into the inner part of the country, after obtaining road permit here. Indians decipher the term meaning 'samudrer jhonkar', indicating the expanse of water nearby in memora-

besides permitting road permits to persons of non-Bhutanese origin. Commerce was mainly in the hands of Indians and the hoteliers were gradually changing to Bhutanese nationality with the aid of Indian merchants. This small town was not as impressive as Phuntsholing.

Kanglung

It was not really an urban centre but may be called an urban settlement as an educational centre with its residential campus grown-up here. The sole college imparting education at degree level — the Sherubtse College — is located here. Its beautiful setting and picturesque buildings, run by missionary ideas and enterprise are really impressive. The centre has a Buddhist temple in the neighbourhood, a rural settlement around it and one of the leading missionary schools not far from it besides a scientifically run agricultural farm.

Tashigang

Tashigang situated in the Tashigang valley (1,067 m or 3,500 ft) is connected with Samdrupjongkhar by road. The township is located on a hill slope, higher than the surrounding valley and is thriving for a reasonably long time. Here the dzong was built in 1668 by the third Deva of Bhutan. Its dzongpons dominated eastern Bhutan in the political history of Bhutan. It was an important cultural centre for the whole of Bhutan in historical past.

Paro

The author's second visit to inner Bhutan in the west was between 1979 and 1984 which coincided with the journey to Paro — an old capital. It was then open to any visitor coming to Thimphu. Paro was then much smaller than the sprawling township of 1989 and onwards when subsequent visits were made. The fertile valley of Paro is quite wide and lies at an elevation of 2,134 m (7,000 ft) east of the town Ha. Actually it is a fast growing settlement next to Thimphu in the west-central Bhutan. Paro dzong was built by Shabdrung Ngawang Namgyal in 1646. Above it is erected Ta Dzong as the sentinal tower. It used to contrcl entrances into Tibet through Drugyal dzong (popularly called by Indians Dugi dzong). A museum has been established recently (in the 1980s) in one of Paro dzongs at higher elevation for tourists' interest and income for the Royal Government, however small it may be.

Ha

The author had an opportunity to visit Ha in one of her subsequent visits to western Bhutan. The settlement is situated at a high elevation of 2,642 m (8,700 ft) in the extreme west, on almost a flat surface. It has an impressive royal palace, a new dzong built in 1915, the older one being destroyed by fire, and a number of public and residential buildings. Tibetans used to enter into Bhutan through this point *via* Chumbi valley. To the Indians Ha is associated with Nehru's first visit to Bhutan.

Punakha

Punakha , an old capital of Bhutan is situated to the north-east of Thimphu at a much lower elevation of 1,463 m (4,800 ft). Its dzong stands at the confluence of two rivers — Pho Chu and Mo Chu which jointly form Chhuzom or Sankosh river. A small iron and wooden bridge leads to the entry to the dzong, which is a massive six-storey building. On way to the dzong a mssionary school building with boarding facility attracts the visitors from outside with its modern architecture. Punakha, on the whole, is a small settlement, an old winter capital founded in 1527 around the dzong built later, perhaps in 1637.

Wangdiphodrang

Wangdiphodrang in west-central Bhutan is situated south of Punakha in the valley of its namesake or Sha. The valley is rich in agriculture and cattle is used for drawing the ploughs as in India. The dzong is situated on the other side of the river connected by an arched-type bridge. The dzong buildings are spread over a few domal shaped hills of different elevations.

Geylegphug

This central Bhutan border town was visited after the trip to western

Bhutan by bus from Phuntsholing to satisfy our curiosity about the area. The settlement is partly inaccessible for the presence of the braided channels or distributories of a large river to be crossed. Our main purpose was to see the township in Chirang and a spa in the interior of central Bhutan. The alignment of the township of Sarbhang reminded me about Aizol in Mizoram. It proved to be very difficult to collect official data for reluctance on their part. On the whole, Geylegphug is well equipped with electricity, water supply, transport facility and official residences. It is another station issuing road permits for the non-nationals for entry into the interior of the country. The hotels are very uncomfortable and the market area is rather unimpressive in appearance.

Tongsa

Tongsa was visited after Wangdiphodrang and Punakha on a different trip, which is located in central Bhutan in the Dangme Chu (Magde chu) valley, a tributary of Tongsa Chu (river). The fortress dzong can be seen again and again from a long distance travelling by car towards Tongsa and Bumthang. The imposing dzong lies at an elevation of 2,286 m (7,500 ft) and a small urban settlement has almost the same elevation, developed on a slightly higher flat with a steep higher hill for the lamas' residence. The linear settlement contains Tibetan hotels and local shops with merchandise for the urban dwellers and the visiting folks and travellers. The dzong was built in 1658 by Shabdrung Ngawang Namgyal and extended by later Penlops. The Tongsa Penlops were powerful in the history of Bhutan and wielded great influence in the political affairs of the country. The present ruling dynasty are the descendants of the Tongsa Penlop, Sir Ugyen Wangchuk who in 1907, became the first hereditary king of Bhutan.

Jakar

The town in Bumthang valley was visited last in 1995 in a package tour. The valley is broad and high. Byakar Dzong lies at an elevation of 2,956 m (9,700 ft) by the bank of Bumthang river which has a round about entry through a majestic gate. The dzong is as impressive as other large dzongs.

In Jakar is situated the famous Jampa Lekhag, one of the earliest monasteries of Bhutan, which we could fortunately visit. It was established in 640 by Tibetan king Srongtsen Gampo converted to Buddhism.

We also visited some old palaces now serving as administrative offices and the palace of the legendary Sindhu Raj, now used by a family of lama. Jakar is not a very small settlement as it is spread over a wide area with various residential and commercial areas having very modern hotels and cultural structures.

Bumthang valley is noteworthy for the bamboo fences made around residences indicating compound boundary which is not observable in other parts of Bhutan in the west. This indicates the region's awareness about the value of land as wealth or property.

NOTES

Bhattacharya, Anima, 1984, 'Demographic, Social and Urban Change in Bhutan', Paper presented at the 25th International Geographical Congress, Paris.

Bhattacharya, Anima 1997, 'Urban Life and Landscape in Bhutan : Some Examples', in *Geographical Thought,* The Department of Geography and Applied Geography, North Bengal University, West Bengal.

Bhutan and Its Natural Resources, 1994, Vikas Publishing House Pvt. Ltd., Delhi.

Bhutan, Country Economic Memorandum, May 1983, Vol I, Main Report.

Druk, Losel, A Quarterly Publication, Royal Govt. of Bhutan (various issues).

Fifth Five Year Plan 1981-1987, Main Document, Royal Govt. of Bhutan.

Kohli, M. 1993, 'Ethnic Crisis in Bhutan', Paper presented at seminar on 'Society and Politics in Bhutan', held in Centre for Himalayan Studies, North Bengal University.

Kuensel (Bhutan's National Newspaper : various issues).

Mehra, G.N., 1978, *Bhutan : Land of Peaceful Dragon,* Vikas Publishing House.

Paramanand, 1993, 'Bhutan's Problems of Ethnicity : Causes, Consequences and Prospects' Paper presented at seminar on 'Society and Politics in Bhutan', held in the Centre for Himalayan Studies, North Bengal University.

Pommaret, F., 1991, 1990, *Bhutan* (translated by Elisabeth B. Booz), Geneva.

Singh, N., 1978, *Bhutan : A Kingdom in the Himalayas,* Thompson Press, New Delhi.

Statistics At A Glance, Central Statistical Organisation, Planning Commission, Bhutan.

Statistical Yearbook of Bhutan 1990, July 1991, Royal Govt. of Bhutan.

World Bank, 1984, *Bhutan : Development in a Himalayan Kingdom,* Washington.

BIBLIOGRAPHY

Aris, Michael, 1980, *The Early History of a Himalayan Kingdom.* Vikas Publishing House Pvt. Ltd., Delhi.

Asian Development Bank, 1998, 'Bhutan' in *Asian Development Outlook.*

Bhattacharya, Anima, 1984, 'Demographic, Social and Urban Change in Bhutan', paper presented at the 25th International Geographical Congress, Paris.

Bhattacharya, Anima, 1997, 'Urban Life and Landscape in Bhutan : Some Examples' in *Geographical Thought,* the Department of Geography and Applied Geography, North Bengal University, West Bengal.

Bohler, W. 1994, 'Has Primary Health Care Reduced Infant Mortality in East Bhutan ? The Effects of Primary Health Care and Birth Spacing on Infant and Child Mortality Patterns in East Bhutan' *Journal of Tropical Pediatrics,* 40 (5), pp. 250-260.

Chhetri, D. B., 1992, 'Forest Fire Season in Bhutan' in *Tsenden : A General Publication on Forestry in Bhutan.* The Department of Forestry, Royal Govt. of Bhutan, Thimphu, Vol. 4, No. 1.

Deb, Arabinda, '*Bhutan — A Study in Frontier Political Relations* (1772-1865), Firma K. L. Mukhopadhyay Ltd., Calcutta.

Dorji, P., 1994, 'The Agricultural Sector Policy : An Emphasis on Conservation and Sustainability' in Sherabtse College, Kanlung, *Bhutan and its Natural Resources,* Vikas Publishing House Pvt. Ltd., Delhi.

Druk Losel, A quarterly publication, Royal Govt. of Bhutan, (Various Issues).

Edmunds, T.O., 1988, *Bhutan — Land of Thunder Dragon,* Elm Tree Books, London.

Geological Survey of India, Bhutan, Bhutan Unit, 1977, *Geology and Mineral Resources of Bhutan.*

Geological Survey of India, Bhutan, Unit, 1979, *Minerals of Bhutan,* Samchi.

Harley, B. and Weber, C. M., 1990, Extended Primary Classrooms in Bhutan,

Report submitted to the Deptt. of Education, Royal Govt. of Bhutan and UNICEF.

Hasrat, J. B., 1980, *History of Bhutan : Land of the Peaceful Dragon,* Education Dept., Royal Govt. of Bhutan.

Ives, J. and Messerli, B. 1989, *The Himalayan Dilemma,* Routledge, London.

Karan, P.P., 1967, *Bhutan — A Physical and Cultural Geography,* University of Kentucky Press, Lexington, U.S.A.

Karan, P.P., 1987, *Bhutan Development amid Environmental and Cultural Preservation,* Institute for the Study of Languages and Cultures of Asia and Africa, Tokyo University of Foreign Affairs.

Kohli, M., 1993, 'Ethnic Crisis in Bhutan', Paper presented at Seminar on *Society and Politics in Bhutan,* held in Centre for Himalayan Studies, North Bengal University.

Kuensal, Bhutan's National newspaper, various issues.

Levitt, M. and Doma, U., 1993, *Motherhood in Bhutan,* Redd Barna and UNICEF.

Mehra, G.N., 1978, *Bhutan Land of Peaceful Dragon,* Vikas Publishing House Pvt. Ltd., Delhi.

Ministry of Agriculture, Royal Govt. of Bhutan, 1993—95, *Agricultural Newsletter,* various issues, nos. 34-45.

Ministry of Agriculture, Royal Govt. of Bhutan, Animal Husbandry Dept., *Projection,* 1961-81.

Ministry of Agriculture, Royal Govt. of Bhutan, Animal Husbandry Dept., *Review of Animal Husbandry Activities in Bhutan.*

Ministry of Agriculture, Royal Govt. of Bhutan, Dept. of Forestry 1993, *Druk Forestry News,* 1, 1 and 1, 2.

Ministry of Agriculture, Royal Govt. of Bhutan, 1993 – 95, *LUP Newsletter,* various issues.

Ministry of Health and Education, Royal Govt. of Bhutan, 1995, *Education Division, General Statistics.*

Morris-Jones, 1985, 'Tales of the Unexpected : The Basic Health Unit in Bhutan' *British Medical Journal,* 291, 28th Sept., 1985, pp. 877-878.

National Environment Commission, 1992, *Common Trees in the Temperate Forest of Bhutan,* Thimphu, Bhutan.

Noble I., 1984, *Bhutan — A Kingdom of the Eastern Himalaya,* Geneva.

Nowrang, R., 1989, 'Medicinal plants and their present status in Bhutan' in *Tsenden : A General publication of Forestry in Bhutan,* the Dept. of Forestry, Royal Govt. of Bhutan, Vol. 1, No. 1.

Olschak, B. C., 1979, *Ancient Bhutan — A Study on Early Buddhism in the Himalaya,* Swiss Foundation for Alpine Research, Zurich.

Paramanand, 1993, 'Bhutan's Problems of Ethnicity, Causes, Consequences and Prospects' Paper presented at Seminar on *Society and Politics in Bhutan,* held in the Centre for Himalayan Studies, North Bengal University.

Planning Commission, Royal Govt. of Bhutan, 1981, *Fifth Five Year Plan,* 1981-87.

Planning Commission, Royal Govt. of Bhutan, 1983, *Bhutan Country Economic Memorandum,* prepared for the Asia and Pacific Round Table Meeting, Geneva.

Planning Commission, Royal Govt. of Bhutan, 1991, *Statistical Yearbook of Bhutan, 1990.*

Planning Commission, Royal Govt. of Bhutan, 1991, *Seventh Five Year Plan 1992-97.*

Planning Commission, Royal Govt. of Bhutan, *Statistics at a Glance,* Central Statistical Organisation.

Pommaret, F., 1991, *Bhutan,* (translated by Elisabeth B. Booz), Geneva.

Raina, B. N., 1994, 'Energy, Environment and Bhutan' in *Bhutan and its Natural Resources,* Sherabtse College, Kanglung, Vikas Publishing House Pvt. Ltd., Delhi.

Ronaldsay, 1977, *Himalayan Bhutan, Sikkim and Tibet,* Ess Ess Publications, Delhi.

Rose, L., 1978, *The Politics of Bhutan,* Cornell U. P., Ithaca.

Rose, L., 1994, 'The Role of Monarchy in the Current Ethnic Conflicts in Bhutan' in Hutt, M. (ed.) *Bhutan : Perspectives on Conflicts and Dissent,* Kiscadale.

Rustomji, N., 1978, *Bhutan — The Dragon Kingdom in Crisis,* Oxford University Press.

Samarasinghe, S., 1990, 'Bhutanese Economy in Transition' *Asian Survey,* pp. 560-575.

Shah, S. 1991 — 'Developing Bhutan's Economy, *Asian Survey,* pp. 816-831.

Singh, N., 1978, *Bhutan : A Kingdom in the Himalayas,* Thompson Press, New Delhi.

Sinha, A. C., 1994, 'Bhutan : Political Culture and National Dilemma' in Hutt, M. (ed.) *Bhutan : Perspectives on Conflict and Dissent,* Kiscadale.

Turner, S., 1800, *A Journey Through Bhutan,* London.

United Nations Development Programme, 1993, *Trust Fund for Environmental Conservation in Bhutan,* Royal Govt. of Bhutan, W.W.F.

UNICEF, 1992, *Children and Women in Bhutan, 1991. A Situation Analysis.*

Ura, K., 1988, *Economic Policy and Development in Bhutan,* Unpublished M. Phil. Thesis, University of Edinburgh.

Ura, K., 1993, *The Nomad's Gamble,* National Environment Commission, Thimphu, Bhutan.

Ura, K., 1994, 'Decentralisation and Development in Medieval and Modern Bhutan' in Aris, M. and Hutt, M., *Bhutan : Aspects of Culture and Development,* Kiscadale.

White, J. C., 1997, *Sikkim and Bhutan,* Vivek Publishing House, Delhi.

Wikan, 1991, *The Girl Child in Bhutan,* UNICEF/University of Oslo.

Wikan and Barth, 1989 — *Bhutan Report, Results of a Fact Finding Mission,* UNICEF.

World Bank, 1984, *Bhutan Development in Himalayan Kingdom,* Washington.

World Bank, 1989, *Bhutan Economic Report,* Washington.

World Bank, 1994, *Bhutan : Country Economic Memorandum,* Washington.

INDEX